7-6-99

Engineering is "the people-serving profession." The work of engineers involves interaction with clients, other engineers, and the public at large. More than any other profession, their work also directly involves and affects the environment. This book makes the case that engineers have special professional obligations to protect and enhance the environment, and the authors – one an engineer and the other a philosopher – seek to provide an ethical basis for these obligations.

The text opens with a series of case studies in which engineers face complex and challenging decisions about the environment. Succeeding chapters examine different ideas about environmental ethics for engineers, including professional codes and both modern and historical discussions of environmental responsibility. The book concludes with a collection of readings that complement the text.

Students, as well as practicing engineers, will find much of interest in this well-argued and thought-provoking book.

Engineering, Ethics, and the Environment

Engineering, Ethics, and the Environment

P. AARNE VESILIND
Duke University

ALASTAIR S. GUNN
University of Waikato

CAMBRIDGE
UNIVERSITY PRESS

PUBLISHED BY THE PRESS SYNDICATE OF THE UNIVERSITY OF CAMBRIDGE
The Pitt Building, Trumpington Street, Cambridge CB2 1RP, United Kingdom

CAMBRIDGE UNIVERSITY PRESS
The Edinburgh Building, Cambridge CB2 2RU, United Kingdom
40 West 20th Street, New York, NY 10011-4211, USA
10 Stamford Road, Oakleigh, Melbourne 3166, Australia

First published 1998

Printed in the United States of America

Typeset in Ehrhardt

Library of Congress Cataloging-in-Publication Data
Vesilind, P. Aarne.
Engineering, ethics, and the environment / P. Aarne Vesilind,
Alastair S. Gunn.
p.   cm.
Includes index.
ISBN 0-521-58112-5 (hc). – ISBN 0-521-58918-5 (pbk.)
1. Engineering ethics.   2. Environmental ethics.   I. Gunn,
Alastair S.   II. Title.
TA157.V42   1997
179'.1'02462 – dc21   97–83
CIP

A catalog record for this book is available from
the British Library

ISBN 0 521 58112 5 hardback
ISBN 0 521 58918 5 paperback

A CONSIDERABLE SPECK

(Microscopic)

A speck that would have been beneath my sight
On any but a paper sheet so white
Set off across what I had written there.
And I had idly poised my pen in air
To stop it with a period of ink,
When something strange about it made me think.
This was no dust speck by my breathing blown,
But unmistakably a living mite
With inclinations it could call its own.
It paused as with suspicion of my pen,
And then came racing wildly on again
To where my manuscript was not yet dry;
Then paused again and either drank or smelt –
With loathing, for again it turned to fly.
Plainly with an intelligence I dealt.
It seemed too tiny to have room for feet,
Yet must have had a set of them complete
To express how much it didn't want to die.
It ran with terror and with cunning crept.
It faltered: I could see it hesitate;
Then in the middle of the open sheet
Cower down in desperation to accept
Whatever I accorded it of fate.
I have none of the tenderer-than-thou
Collectivistic regimenting love
With which the modern world is being swept.
But this poor microscopic item now!
Since it was nothing I knew evil of
I let it lie there till I hope it slept.

I have a mind myself and recognize
Mind when I meet with it in any guise.
No one can know how glad I am to find
On any sheet the least display of mind.

– Robert Frost

Contents

Preface

An ethical analysis produces no absolute answers. This can be quite disconcerting because we engineers expect valid, correct, and useful answers to problems. When we study statics, for example, we all know what the rules are and we can all agree on the right answers. In ethics, however, the best we can do is argue that some answers are better than others, and, of course, these answers are always open to disagreement.

Engineers, as professionals, have a special responsibility to the public, and this responsibility is often expressed in terms of professional ethics. Engineers invariably face situations where values become variables in the decision-making process. Indeed, the ethical aspects of a decision often prove more difficult than the technical. Recognizing this, the engineering profession has strongly encouraged engineering schools to introduce more professional ethics into engineering curricula. Professional ethics has rightly become an integral part of engineering education.

Professional engineering ethics relates to how engineers, in their professional roles, interact with other people – clients, other engineers, or the public in general. These questions can be complex and the dilemmas difficult.

Engineering ethics is tricky enough when it concerns only how engineers relate to each other and the public; it becomes trickier still when we also consider how engineers ought to react to the non-human environment. That is, when we ask what is the engineers' environmental ethic?

Environmental ethics, to an even greater degree than ordinary ethics, is a subject without definition and without consensus. And yet, every person on this planet makes everyday decisions that relate to environmental ethics. Questions as simple as "What should I eat?" or "How should I move from place to place?" all raise environmental and ethical issues.

Environmental ethics is especially important for engineers, because so much of their work affects the environment. How should the engineer balance the human gains of development against environmental damage? When should the engineer maintain client confidentiality in the face of potential environmental problems?

Not all practicing engineers agree that understanding environmental ethics is worth the effort. They might consider many of the scenarios described in this book as irrelevant to their professional objectives, or they might deny that there could be more than one acceptable answer to a value-laden problem. These engineers are, of course, simply displaying their own values, such as loyalty and job security – values that the general public or other engineers might weigh differently.

Reading this book will not necessarily make anyone a better engineer, or a better person. The reader may find that there are legitimate competing values and discover that decision making in engineering is not as straightforward as it may often appear to the layperson.

Likewise, engineering students who study this book as a text will not be magically transformed into environmental saints. We all carry too many pre-conditioned and prejudiced ideas around with us as part of our psychological baggage to be so easily converted. And the intent of the authors is certainly not to convert anyone to a particular ideology. We do hope, however, that if this book is used in the classroom, spirited discussions of environmental ethics will enhance the decision-making skills of young engineering students so that when they enter the working world they will better understand the issues and difficulties inherent in making decisions concerning the environment. If this book is read by interested readers, we hope they will attain a new appreciation for the complexity of the problem and will be able to approach environmental questions with an enhanced understanding and a renewed perspective.

We must emphasize that this book does not contain a list of detailed and specific principles that, if followed, will lead to environmentally sound behavior. As we note in the final chapter, an exhaustive list of environmentally sound principles simply does not exist. Environmental ethics is very recent as an academic discipline and there is much to be done.

The need for an environmental ethic is, however, critical. The earth's climate is already changing measurably, the rate of human population increase appears to be out of control, and no part of the earth is untouched by human activity. If sound engineering practices, in the widest sense, are to become the norm, a functioning environmental ethic, perhaps based on different foundations, is absolutely necessary. The intent of this book is to contribute toward the formulation of that ethic.

Some of the material, in earlier versions, appeared in *Environmental Ethics for Engineers,* published in 1986 by Lewis Publishers. The present book was written while the senior author was on a National Science Foundation Fellowship at Dartmouth College. Appreciation is expressed to both the NSF Ethics and Values Studies program, Dr. Rachelle Hollander, Director, and the Institute for Applied Ethics at Dartmouth, Dr. Deni Elliott, Director. Pamela Vesilind, representing the next generation, typed much of the early work on this book, and challenged the senior author to defend his views of life. Thanks to her on both counts. Another member of the next generation, Christopher Endy, edited the final draft, and many of his suggestions were incorporated into the text. The book in manuscript form was "beta tested" by Dr. Joseph Delfino of the University of Florida, and the comments of his students are very much appreciated.

This book is dedicated to Carole Gunn, who would not even think of running over a box turtle, and to Libby Vesilind, who hugs trees.

PAV
ASG
1997

A Note on Terminology

Some ethics texts pay a great deal of attention to the definition and analysis of terms such as "ought," "right," "good," "wrong," "ethics," and "morality." These are fundamental terms in any inquiry into how we should live and what values we share. Texts on ethical theory often analyze the many uses of these words in ordinary language and argue for one usage or another.

While in mathematics, logic, and philosophy, exact meanings are necessary for understanding, this is not true in other fields such as engineering. For engineers, and for most people, ordinary words often suffice since their definition is evident through the context. For example, "good" is a very general term of commendation, and a good engineer, a good life, a good haircut, good sewers, good luck, and a good question are "good" in different senses, but everyone can figure out which is the appropriate sense because of the noun the word modifies. In this book, we use language in its ordinary sense, and if there are various senses in which a word may be understood, we trust that the reader can understand the sense from the context.

However, to avoid undesirable ambiguity, we need to explain our use of "ethics," "morality," and their derivative words.

Morality refers to value-laden questions that arise at a personal level but relate to the functioning of human societies. That is, moral questions are questions about how one ought to live as a citizen or as a person. Morality relates to questions of doing the right thing. Much like religious belief, morality cannot be "taught" through reasoned arguments. In other words, nobody can use reason to make us believe that we should not steal from others, or that we should believe in a certain religion. We are, of course, "taught" morality (and religion) in the sense that our parents, or teachers, or coaches, or peers show by example or command that certain things are right and others wrong. We begin to believe these principles, but not necessarily because of rational argument.

Ethics, on the other hand, is the systematic analysis of morality. Ethical analysis can be theorized and systematized, and just like other theories, can be taught through reason and logic. We can learn certain principles or ethical theories, and apply these to different practical situations. The various theories may, however, provide different suggested courses of action. Although what we ultimately do depends on our sense of morality, the process of ethical reasoning allows us to understand better our values and the ramifications of our actions.

The word "ethics" can also take on modifiers, and in such cases the term means the systematic analysis of morality in a particular field or profession. For example, environmental ethics concerns relations between humans and the rest of nature. It differs from other types of ethics because of its subject matter. Professional ethics such as engineering (or medical or legal) ethics concerns role-related behavior; it is the behavior of persons in their roles as engineers (or doctors or lawyers).

While professional ethics such as engineering ethics is meaningful only to the profession (except as the well-being of the public is affected), environmental ethics applies to every human being. Because engineers so often interact with the environment in the course of their professional duties, they must consider both types of ethics – professional ethics and environmental ethics. For this reason, environmental ethics for engineers takes on greater meaning and demands deeper commitment.

Part I

1

The Problem of Environmental Ethics in Engineering

"It's not easy being green," laments Kermit the Frog. We could also echo, "It's not easy being an engineer – let alone a *green* engineer!"

Engineering, as one of the professions, is meant to serve the public. The profession requires that its practitioners adhere to strict codes of conduct, such as at all times placing "paramount the health, safety and welfare of the public" that the engineer serves.[1]

While this seems fairly straightforward, it also raises troubling questions, such as:

- Just who is the public? Does it include everyone who might be affected by a decision, everyone who has an opinion about an issue?
- Does the public include future generations, and if so, how far into the future?
- Are engineers in a position to know what is best for the public, and if so how should they decide this?
- Do engineers have managerial as well as technical responsibilities?
- What level of safety should the public expect; what is "acceptable risk"?

Thus, even when engineering ethics rests on the human welfare principle, it is still complex and full of potential conflicts.

When we introduce the environment as an additional responsibility of the engineer, we add another dimension to the puzzle. As long as the environment can be thought of as existing only for the benefit of present humans (which we refer to as "the traditional approach" or "traditional ethics"), environmental concerns fit nicely with the human welfare principle. But when environmental concerns spill over into animal welfare, endangered

---

1. Code of Ethics, American Society of Civil Engineers, "Fundamental Canons," item 1.

species, and the preservation of natural systems, the traditional approach to ethics seems to be inadequate.

Engineers, unlike the other professions, are directly involved in environmental conservation and preservation. No matter what the project, engineers are the ones who will *do* it. To build a dam requires the skills of many professionals, such as accountants, lawyers, and geologists. But it is the engineer who actually builds it, and because of this, the engineer has a special responsibility toward the environment. In short, the engineer can *make a difference.*

Because of this responsibility, modern engineering takes on a special complexity. Engineers can find themselves in unexpected situations they might be ill-prepared to handle. The following are some scenarios in case study form that illustrate the complexity of modern engineering. While you read them, ask how you would act in such a situation. Then keep these case studies in mind as you read the other chapters.

Case 1:   Now Where Did I Put That Stuff?

Peter has been working with Bigness Oil Company's local affiliate for several years and has established a strong, trusting relationship with Jesse, the manager of the local facility. The local company receives various petrochemical products via pipeline and tank truck and blends them for resale to the private sector. The facility, on Peter's recommendation, follows all environmental regulations to the letter, and has a solid reputation with the state regulatory agency.

Jesse has been so pleased with Peter's work, he has recommended that Peter be retained as the corporate consulting engineer. This would be a significant advancement for Peter and his consulting firm. There is talk of a vice-presidency for Peter in a few years.

One day, over a cup of coffee, Jesse tells Peter an old story about a mysterious loss in one of the raw petrochemicals received by pipeline. Sometime during the 1950s, when operations were pretty lax, 10,000 gallons of one of the process chemicals were discovered missing when the books were audited. After running pressure tests on the pipelines, the plant manager found that one of the pipes had corroded and had been leaking the chemical into the ground. After stopping the leak, the company sank observation and sampling wells and found that the product was sitting in a vertical plume, slowly diffusing into a deep aquifer. Because there was no surface or groundwater pollution off the plant property, the plant manager decided to do nothing. The wells were capped, and the story never appeared in the press. Jesse thought that somewhere under the plant the plume still existed, slowly diffusing into the aquifer, although the last tests from the sampling wells showed that the concentration of the chemical in the groundwater within 100 meters of the surface was essentially zero.

Peter is taken aback by this apparently innocent revelation. He recognizes that state law requires him to report all spills, but what about spills that occurred years ago and seem to have dissipated already? He frowns and says to Jesse, "We have to report this spill to the state, you know."

Jesse is incredulous.

"But there *is* no spill. If the state made us look for it, we probably couldn't find it, and even if we did, it makes no sense whatever to pump it out or contain it in any way."

"But the law says that we have to report . . ." begins Peter.

"Hey look. I told you this in confidence. Your own engineering code of ethics requires client confidentiality. And what would be the good of going to the state? There's nothing to be done. The only thing that would happen is that the company would get into trouble and have to spend useless dollars to correct a situation that cannot be corrected and does not need remediation."

"But . . ."

"Peter. Let me be frank. If you go to the state with this, you won't be doing anyone any good – not the company, not the environment, and certainly not your own career. I cannot have a consulting engineer who does not value client loyalty."

Should Peter report the spill to the state?

Case 2:   But It Really Is Best for All Concerned, Isn't It?[2]

Jason is the newest member of an environmental engineering firm in a small southwestern city. The firm is the newest and smallest in the city and has had difficulty obtaining work. The situation has become so desperate that the company is struggling for survival and has been extending itself beyond its resources to avoid layoffs. The other firms are not hiring, and given the seniority policy of the company, Jason realizes he would be the first to be laid

---

2.   This case study is based on a similar exercise in Joan Callahan's book *Ethical Issues in Professional Life* (Oxford University Press, New York, 1988).

off. Jason has recently moved to the community because of his wife's career opportunities, and they have decided to settle permanently there.

Being new to the area, Jason is appalled at the dismal sanitary conditions in some of the poorer parts of the city, especially the absence of a sewerage system. The failed seepage fields of the on-site wastewater disposal systems are clearly creating a public health hazard. Much of the outlying region depends on groundwater for a drinking water supply, and Jason recognizes that the groundwater is being steadily and irreversibly polluted by the inadequate waste disposal.

The city is now considering installing a sewerage system in the poorer neighborhoods, and Jason is asked to write a technical and cost proposal for the firm in response to the city's request for proposals. After submitting his draft to Michele, the firm's president, Jason is called into her office.

"I need you to cut this proposal by roughly $300,000," says Michele.

"No way. We've already cut as much as possible. We'd lose our shirt," counters Jason.

"Listen to me, Jason. I've spoken to the mayor, who, as you well know, is currently in the last stages of his campaign for reelection. He needs funds in order to bring out the vote. You've heard the mayor's opponent. She's been opposed to the sewerage system because, she says, building and maintaining it would necessitate an increase in city taxes. She's running on a 'cut taxes' platform and if she wins, there will be no sewerage system. The timing is critical. Whoever wins will decide on the project."

Michele tells Jason that after much painful thought she has privately arranged for a large contribution (adequately laundered) to reach the mayor's campaign fund. In exchange, the mayor has instructed her how to write the proposal so it will be accepted by the city board with no challenge. Both Michele and the mayor know there will be cost overruns after the project has begun, but they have agreed to a mechanism for covering these additional costs. The mayor knows that if the firm does the project, it will be done well and will not cost the city much more than if one of the other firms handles it. The mayor also strongly believes the city is better off with as many engineering firms as possible, and thus wants to see the firm stay in business.

"If you refuse to alter the proposal," Michele tells Jason, "then I will do it myself," with the unstated understanding that Jason can then expect to be laid off. If the firm does not get the contract, there is every reason to believe not only that Jason will lose his job, but that the firm will soon go out of business. If the incumbent mayor loses, there will be no sewerage system at all and the deplorable environmental conditions will continue.

Should Jason go along with the plan and revise the proposal?

Case 3: You've Seen One Big Tree, You've Seen Them All

Libby was in good spirits. She loved her job as assistant engineer for the town, and the weather that day was perfect for working out-of-doors. She had found a job as the on-site inspector for the new gravity trunk sewer. The job gave her significant responsibility and even allowed her to get out of the office. Not bad for a young engineer only a few months out of school.

The town was doing the work in-house, partly because of Bud, an experienced foreman who would see that the job got done right. Bud was wonderful to work with and was full of stories and construction know-how. Libby expected to learn a lot from Bud.

This particular morning the job required cutting and clearing a strip of woods on the right-of-way. When Libby arrived, the crew was already noisily getting prepared for the morning's work. She decided to walk ahead up the right-of-way to see what the terrain was like.

About a hundred yards up, she came upon a huge oak tree, somewhat off the centerline but still on the right-of-way, and therefore destined for cutting down. It was a magnificent oak, perhaps 300 years old, which had somehow survived the clear-cutting on this land in the mid-1800s. Hardly any other trees there were over 150 years old, all having fallen to the tobacco farmer's thirst for more land. But here was this magnificent tree. Awesome!

Libby literally dragged Bud up to it and exclaimed, "We can't cut down this tree. We can run the line around it and still stay in the right-of-way."

"Nope. It has to come down," responded Bud. "First, we are running a gravity sewer. You just don't change sewer alignment. We'd have to construct additional manholes and redesign the whole line. And most importantly, you can't have such a large tree on a sewer line right-of-way. The roots will eventually break into the pipe and cause cracks. In the worst case, the roots will fill up the whole pipe, and that requires a cleaning and possible replacement. We simply cannot allow this tree to remain here."

"But think of this tree as a treasure. It's maybe three hundred years old. There probably are no other trees like this in the county," implored Libby.

"A tree is a tree. We're in the business of building a sewer line, and the tree is in the way," Bud insisted.

"Well, I think this tree is special, and I insist that we save it. Since I am the engineer in charge"—she gulped inwardly, surprised at her own assertiveness—"I say we do not cut down this tree."

She looked around and saw that some of the crew had walked up to them and were standing around, chainsaws in hand, with wry smiles on their faces. Bud was looking very uncomfortable.

"Okay," he said. "You're the boss. The tree stays."

That afternoon in her office, Libby reflected on the confrontation, and tried to understand her strong feelings for the old tree. What caused her to want to save a tree? So what if it was special? There were many other trees that were being killed in order to run the sewer line. What was special about this one? Was it just its age, or was there something more?

The next morning Libby went back to the construction site and was shocked to find the old tree gone. She stormed into Bud's construction trailer and shouted, "Bud! What happened to the old tree?"

"Don't you get your pretty head upset now. I called the Director of Public Works and described to him what we talked about, and he said to cut down the tree. It was the right thing to do. If you don't agree, you have a lot to learn about construction."

What are Libby's alternatives? What would *you* do?

Case 4: Free Enterprise at Work

One of the reasons Nathan had moved to the rural community in Montana was to get away from the pressure of city life. What he didn't count on was the difficulty of relocating his engineering practice. He'd had a small but profitable firm in Cleveland, but had decided to sell out and move so he could enjoy his family and his profession without the pressures of keeping the office active. Now he was finding that without his network of friends and clients, getting work was difficult. Several small local firms could do the same type of general environmental engineering work that Nathan could do, and they all had good reputations with the local clientele.

It was therefore with some relief and excitement that Nathan received a call one morning from a prospective client who would not discuss his needs over the telephone but wanted Nathan to drive out to his ranch. It didn't take Nathan long to decide to make the trip, even though the ranch was over an hour away. The directions led him into the mountains and onto a dirt road that snaked its way through a canyon, suddenly opening up into a large, secluded valley. At the only house in sight he met his prospective client, a wealthy local rancher named Wayne.

"I won't waste your time," Wayne started right in. "I have big plans for this place and I need your engineering skills to help me do it. I and several of my colleagues are planning to build an exotic game ranch here, buying live animals like tigers, lions, and polar bears and setting them loose inside a fenced-in area. Hunters who've always wanted to bag such game will come and shoot them. We will arrange the whole hunt, make sure they bag what they contracted for, and mount the heads for them as trophies. There are people out there who are willing to pay very well indeed to get their tiger.

"What I need, Nathan," continued Wayne "is for you to be my engineer for the entire construction phase. We'll be building fences, holding pens for the animals, a clubhouse, shooting shed, and all of the water and wastewater systems. I've allotted ten million for the construction. Your construction management fee will be about two hundred thousand. I want you because I hear you're a darn good engineer. I also understand that you're new in town and wouldn't go around blabbing about this to everyone, causing adverse publicity and getting the eastern press all in an uproar. What do you say?"

Nathan was taken aback. He had heard of such ranches but never thought he'd be asked to be the construction engineer for one. He finally inquired, rather meekly, if the ranch was legal.

"Of course it's legal," replied Wayne. "We'll be buying the animals from trappers all over the world, and shipping them in."

"What about endangered species?" asked Nathan.

"No problem. We'll only use animals that are not on the endangered species list," Wayne assured, adding with a wink, "But if our suppliers happen to make a mistake, we certainly wouldn't offend them by rejecting the shipment, would we?

"Look, it's essentially on the up-and-up. We simply bring in some animals and shoot them. What's the big deal? People have been killing animals since the caveman days. And your involvement here is only during the construction phase. If you're not a sportsman, no problem. You just do the construction supervision for us, and get well paid. You don't have to be here once we start operating.

"So do we have a deal?" asked Wayne.

"I don't know. Give me a little time to think about it," Nathan replied.

"Sure. Take as much time as you need. But tell me in the next ten minutes. I'm a busy man."

What should Nathan do?

Case 5:   The Magic Bullet

The state had promised to come down hard on Domestic Imports, Inc., if it violated its discharge permit one more time, and Sue was on the spot. It was her job to operate the wastewater treatment works, but the firm's management had rejected all her requests for improvements and expansion. Management believed the treatment works had sufficient capacity to deal with the waste. This was true on paper if the average flow was used in the calculation. Unfortunately, the manufacturing operation was such that slug flows could come unexpectedly and the biological system simply could not adapt fast enough. The effluent BOD[3] shot sky-high for a few days and then settled back again to below acceptable discharge standards. If the state showed up on one of the days during the upset, Sue was in big trouble.

One day Sue is having lunch with a friend, Emmett, who works in the quality control lab. She explains to Emmett that she needs to figure out how to reduce the upsets when the slug loads come in. She's already talked to the plant manager, to get him to assure her that the slug loads won't occur, or to build an equalization basin, but both requests were denied.

"The slug load is not your problem," Emmett suggests. "It's the high BOD that results from the slug load upsetting your plant."

"Okay, wiseguy. You're right. But that doesn't help me."

"Well, maybe I can. Your problem is that you're running high BOD and you need to reduce it. Suppose I told you that you could reduce the BOD simply by bleeding a chemical additive into the line, and that you can add as much of this chemical as you need in order to reduce your BOD. Would you buy it?" Emmett asks.

"Just for fun, suppose I would. What is this chemical?"

"I've been playing around with a family of chemicals that slows down microbial metabolism but doesn't kill the microbes. If you add this stuff to your line following the final clarifier, your effluent BOD will be reduced because the metabolic activity of the microbes will be reduced. This means they will use oxygen at a slower rate. You can set up a small can of it and whenever you see one of those slugs coming down from manufacturing you

---

3.   BOD is Biochemical Oxygen Demand, a measure of the amount of oxygen needed by microorganisms while decomposing organic material. The purpose of most wastewater treatment plants is to reduce the BOD of the wastewater so the oxygen levels in the streams will not be depressed below some critical level that will support fish life.

start to bleed in this chemical. Your BOD will stay within the effluent limits, and in a few days when things have calmed down, you turn it off. Even if the state comes to visit, there's no way they can detect it. You're doing nothing illegal. You're simply slowing down the metabolic activity."

"But is this stuff toxic?" asks Sue.

"No, not at all. It shows no detrimental affect in bioassay tests. You want to try it out?"

"Wait a minute. This is complicated. If we use this magic bullet of yours, our BOD will be depressed, we pass the state inspection, but we haven't treated the wastewater. The oxygen demand will still occur in the stream."

"Yes, but many miles and many days downstream. They will never be able to associate your discharge with a fish kill, if in fact this would occur. What do you say? Want to give it a try?"

Should Sue use the chemical?

Case 6: Just a Minor Editorial Change

Stan works as an engineer in the Atlanta office of a large consulting firm. His current job is to write an environmental impact statement for a neighborhood association. The study concludes that the plans by Megatane Oil Company to build a petrochemical complex would harm the habitat of several endangered species. The client, the neighborhood association, has already reviewed draft copies of Stan's report and is planning to hold a press conference when the final report is delivered. The association asks Stan to attend the news conference in his professional capacity.

A few days before the conference, Bruce, also an engineer and a partner in the firm, receives a phone call in the New York office.

"Bruce, this is J. C. Octane, president of Megatane Oil Company. As you well know, we have retained your firm for all of our business and have been quite satisfied with your work. We intend to build a refinery in the Atlanta area, and hope to use you as the design engineers."

"We'd be pleased to work with you again," replies Bruce, already counting the million-dollar design fee.

"There is, however, a small problem," continues J. C. Octane, "It seems that one of your engineers in the Atlanta office has conducted a study for a neighborhood group opposing our refinery. I've received a draft copy of the study, and my understanding is that the engineer and leaders of the neighborhood organization will hold a press conference in a few days to publicize the alleged unfavorable environmental impact as a result of the refinery. I needn't tell you how disappointed we'll be if this occurs."

As soon as Bruce hangs up the phone with J. C. Octane, he calls Stan in Atlanta.

"Stan, you must postpone the press conference at all costs," Bruce yells.

"Why? It's all ready to go."

"Here's why. You had no way of knowing this, but Megatane Oil is one of the firm's most valued clients, and the president of the company has found out about your report and threatens to pull all of their business if the report's delivered to the neighborhood association. You have to rewrite the report in a way to show there would be no significant damage to the environment."

"I can't do that!" Stan pleads.

"Let me see if I can make it clear to you, then," replies Bruce. "You either rewrite the report or withdraw from the project and write a letter to the neighborhood association stating that the draft report was in error and offering to refund all of their money. You have no other choices!"

Does Stan have another choice? What should he do?

Case 7:   The Gaia Sacrifice

James is the newly appointed project manager of a geothermal power station being built in the nation of Terranova by the New Zealand construction company that employed him three years earlier. Construction work is well under way, and the station is to be commissioned in six months time. James is pleased and surprised by the chance to work on the project. The original project manager, Adam, with considerably more experience, had been very enthusiastic about the project but apparently left at short notice.

Soon after arriving in the town of Opra, where the project is based, James has dinner with Laura, another expatriate New Zealander, who has been living in Terranova for several years. As a new arrival, James is keen to learn about the local culture, and Laura is happy to oblige. Among other things, he learns that the local people, followers of the doctrine of Gaia, believe that all the Earth is one personal organism named Gala. As Gaians, they must cherish and respect the Earth.

"That must make it hard for them to go along with projects like this power station," James says. "I mean, how can they accept making a hole *that* big in Gaia and then sucking all the goodness out of it?"

"We all wondered about that," Laura responds, "and evidently they did too. The Terranovan government didn't really consult them, but they eventually learned about the project and had a meeting of all the people from the region. Adam was involved because as project manager he knew the cooperation of the locals was important to the project. He was also interested in how they'd deal with it. Did you know he had a degree in anthropology before he did his engineering training?

"Anyway, this is what happened. The elders decided that although large construction projects are not part of their tradition, this one will bring great benefits to the people – electricity, jobs, a better life – provided Gaia's permission is obtained."

"But how will they know that?" James asks.

"Yes, well, now we come to the morbid bit. The elders decided that Gaia's approval would depend on the sacrifice of the first baby born in Opra after the plant is completed, to make up for the injury done to Gaia by the earthworks."

"But that's dreadful!" James explodes. "How can they sacrifice an innocent baby for the sake of some superstition?"

"That's just what Adam said to me," Laura replies. "And he's an anthropologist! I said, 'Look Adam, these people have their own customs and it's not up to you to judge. I expect they think your religion is superstitious nonsense, too. So do I, if it comes to that, but that's your choice.'"

"But surely human sacrifice is illegal in Terranova!" James exclaims.

"Technically, yes, but the central government prefers not to interfere in local traditions, as Adam found when he complained to the regional official of the Ministry of Development. I'll give Adam credit, he did try to change things. He argued with the elders, tried to get them to accept an animal sacrifice, even offered to build a temple to Gaia if they'd do without the sacrifice. But there was no way around it. They've made it quite clear: if the ceremony isn't carried out, the plant can't be commissioned and nobody will be allowed to enter it because it will be cursed by Gaia. By the way, since obviously nobody's told you, they also wanted Adam to attend because he was the man in charge. In fact they even wanted him to wield the sacrificial knife at first. When he started objecting that it was against his religion, they agreed to let him off, but they still wanted him to be there. And that's why he left. I don't want to be unkind, but that's why you're here."

"Didn't he talk to his boss back in New Zealand?" James wondered. "Surely they would have backed him up."

"Yes, he did, and I heard the conversation because Adam recorded it and played it back to me. Adam told his boss that even if the authorities in Ter-

ranova look the other way at human sacrifice, it's seriously illegal in New
Zealand. His boss responded that New Zealand law doesn't apply in Terranova, and that nobody back home will ever find out about it. His boss said it
was his job to deal with local customs – that's what he's paid for – and the
important thing to remember is that the firm is in business to make a profit,
and it won't be paid for a power station where nobody will work."

"There must be something someone can do about this!" James says.
"What can *I* do about it?"

"You want my advice? Just stop moralizing and do whatever needs to be
done to get this project finished."

Should James just get on with it?

Case 8:   Just Change the Drilling Log

Jaan, an engineer fresh out of school, is busily at work at a local consulting
firm. As one of his first jobs, he must oversee the operation of a hazardous
waste cleanup at a Superfund site. The remediation plan is to drill a series of
interception wells and capture the contaminated groundwater. For safety,
each well must reach bedrock.

One day, the foreman calls Jaan over to a well being drilled.

"Strange," says the foreman, "We were supposed to hit bedrock at 230
feet, and we are already at 270 and haven't hit anything yet. You want me to
keep going?"

Not knowing exactly how to respond, Jaan calls the office and talks to
Robert, the engineer who is his immediate superior.

"What do you mean we haven't hit rock?" inquires Robert, "We have all
the borings to show it's at 230 feet."

"Well, the foreman says he hasn't hit anything yet," replies Jaan. "What
do you want me to do?"

"We are working on a contract, and we were expecting to hit rock here.
Maybe the drill has bent, or we're in some kind of seam in the rock. Whatever, we can't afford to keep on drilling. Let's just stop here, show on the
drilling log that we hit rock at 270 feet. Nobody will ever know."

"We can't do that! Suppose there *is* a seam down there? The hazardous
waste could seep out of the containment."

"So what. It'll be years, maybe decades, before anyone will know. And
besides, it probably will be so diluted by the groundwater that nobody will

even be able to detect it. Just note on the drilling log that you hit rock, and move to the next site."

Should Jaan just do what he is told?

Case 9:   You'll Never Know It's There

Diane works as an engineer for a large international consulting firm that has been retained by the Arizona government to assist in the construction of a natural gas pipeline. Her job is to lay out the centerline of the pipeline according to the plans developed in Washington.

After a few weeks on the job, she is approached by the leaders of a local Navajo village and told that the gas line might traverse a sacred Navajo burial ground. She looks at the map and explains to the Navajo leaders that the initial land survey did not identify any such burial grounds.

"Yes, although the burial ground has not been used recently, our people believe that in ancient times this was a burial ground, even though we cannot prove it. What matters is not that we can show by archeological digs that this was indeed a burial ground, but rather that the people believe it was. We therefore would like to change the alignment of the pipeline to avoid the mountain."

"I can't do that by myself. I'd have to get approval from Washington. And whatever is done would cost a great deal of money. I suggest that you not pursue this further," Diane replies.

"We already have talked to the people in Washington, and as you say, they insist they cannot accept the presence of the burial ground in the absence of archeological proof. Yet to our people the land is sacred. We would like you to try one more time to divert the pipeline."

"But the pipeline will be buried. Once construction is complete, the vegetation will be restored, and you'll never know the pipeline is there," suggests Diane.

"Oh, yes. We will know it is there. And our ancestors will know it is there."

The next day Diane gets on the telephone with Tom, her boss in Washington, and tells him about the Navajo visit.

"Ignore them," Tom advises.

"I can't ignore them. They truly *feel* violated by the pipeline on their sacred land," replies Diane.

"If you're going to be so sensitive to every whim and wish of every pressure group," Tom suggests, "maybe you shouldn't be on this job."

What are Diane's alternatives?

Case 10:   Running Over Box Turtles

Carole was late for work, and she was stepping on it, scooting down the rural highway. Her mind was on the problems she was having with the state wildlife people. They kept talking about "ecosystems" and "habitat" and "endangered species." And all *she* wanted to do was build the dam for her client, a reservoir that would bring a new source of much needed water to the community. Its citizens were paying good money for this reservoir. Why the state people kept wrangling about wildlife was a mystery to Carole. Didn't they know what was important?

As she kept her eyes on the road and her mind on the job, she spotted a dark brown blob in the middle of the road. As she got closer, she saw that it was a small box turtle, trying to make it across the road. She swerved her car to avoid hitting it, and saw in her rearview mirror that although she might have scared the wits out of the turtle, it seemed to have survived the experience and was continuing on its way. She smiled to herself and felt pleased that she had not run over the box turtle. It seemed to be the right thing to do.

She was pleased – until she started to think about why she had avoided hitting the turtle. Of what use was it to her anyway? But instead she had swerved her car, risking an accident, in order to avoid hitting the turtle. Not very rational. Not very bright. She had gone to school all these years, had ten years of engineering experience, earned her professional engineering license, and was a respected member of the engineering community. And she had instinctively avoided killing a stupid box turtle while endangering her own life and perhaps the lives of other people.

"Okay," she said to herself. "Let's figure this out. This is a neat problem to think about. Certainly it's a better way to spend the time than stewing about the stupid state wildlife people."

So she began to construct arguments to explain why she did what she did, and why she thought in retrospect that this was in fact the right thing to do. But to her consternation, she could not come up with a single logical argument as to why she *ought* to have spared the turtle. She realized that if she could not develop a logical argument as to why she ought to avoid hitting the turtle, she would never be able to convince others to act in a similar way.

She arrived at work totally frustrated with her inability to explain her actions, much less to present a cogent argument for recommending similar action to others. But the frustration soon changed when she saw another letter from state wildlife people proposing yet another delay in the reservoir project. They wanted to do some more studies on the destruction of habitat for the Venus's-flytrap, an endangered species. She read the letter and stormed out of the office, ready to do battle with the state agency on behalf of her clients.

What are Carole's conflicts? Why does she feel frustrated about the turtle incident? Why does she seem to care less about the Venus's-flytrap, an endangered species, than a box turtle on the highway?

Suggested Supplemental Readings for Chapter 1

1. *The Kepone Tragedy*, W. Goldfarb
2. *The Hooker Memos*, CBS News
3. The *Bunker Hill Lead Smelter*, C. Tate

2

Engineering as a People-Serving Profession

Carole's problem in Case 10 is common among engineers. She has a sense of the worth of the natural environment, but she also recognizes that she has a professional responsibility to her clients. Like most engineers, she has difficulty reconciling her attitude toward nature with her job. What she probably does not realize is that her difficulty arises from the integration of her attitude toward non-human nature and the commitment she feels toward her professional work.

Engineers like Carole are often called upon to make value-laden decisions, and how these decisions are made often affects both their own careers and the public good. In this chapter, we first explore the moral development of engineers, then discuss how engineers view their own profession and, in contrast, the public's view of engineering. We then ask what responsibilities engineers have to distant peoples and to future generations. Finally, the idea of environmental ethics in engineering is introduced.

2.1   Engineering as a Profession

While all professions have had their admirers, they have also often been mistrusted by the public. As George Bernard Shaw put it, "All professions are a conspiracy against the laity."[1] To their critics, professionals restrict entry to training and registration, suppress new ideas in order to protect their oligopolistic control, and use obfuscation and paternalism to bolster their authenticity.

We do not deny that professionals gain considerable benefits – prestige, independence, and financial rewards – from their status, or that professional

---

1.   G. B. Shaw, "The Doctor's Dilemma," *Six Plays by Bernard Shaw* (Dodd, Mead, New York, 1941), from the Preface.

societies have sometimes ignored aspects of the public interest in order to protect the interests of their members. We also accept that professions are usually conservative, and that sometimes this conservatism has not been in the best interest of consumers and clients. However, we strongly deny that the career benefits of professionals come at the expense of society. Sound professional practice, conducted in a responsible manner by trained specialists, is in everyone's interest.

Although our discussion in this chapter focuses on the behavior of individual engineers, many engineers are employees of corporations. Questions of corporate ethics are out of the scope of this book, but we should point out that we are not claiming that corporate ethics is nothing more than the ethical behavior of individual employees. There is a difference between *an ethical organization* and *an organization of ethical people*. Notoriously, good people can do awful things in their jobs, like the German doctors who "euthanized" children classed as "defective" under the Nazi regime. Conversely, an organization could be highly ethical, providing useful goods and services at reasonable prices, treating its employees fairly, and being a good corporate citizen. Yet its employees could all be cynical and selfish people who have learned how to act in ethically acceptable ways at work, but whose sole interest in their jobs is to make lots of money and enjoy the excellent conditions provided by the firm, and who in their private lives are vicious and cruel.

In other words, the properties of an organization will not necessarily be those of its members, and vice versa. Analogously, a great football team does not have to consist entirely of great players, nor do eleven great players necessarily make a great team. Still, these are probably untypical cases. That we find the behavior of the Nazi doctors so puzzling suggests that we do normally expect a person's public and private behavior to be governed by the same standards; and we would surely not trust the cynical employees to act ethically except when there is something in it for them.

Engineering is, however, more than just employment. Irrespective of whether engineers are self-employed, work for the public sector, or are employees of corporations, we maintain that engineering *professionalism* is in the public interest; and we also believe that most engineers are motivated by a genuine and principled concern for the public good.

But how do such principles develop? How is it that engineers by and large *do* place the health, safety, and welfare of the public paramount in what they do? One possible explanation may lie in a theory known as "moral development."

2.2   Moral Development and Professional Engineering

Each person during his/her life and career develops moral standards. In this section we examine two aspects of these standards. First is the question of moral psychology, which asks how people come to hold their ethical standards. Second – the main subject of the book – is the issue of how people decide what is a basis for values. One explanation for such questions is the idea that "normal moral development" parallels normal physical or intellectual development. In this view, humans are organisms that in an appropriate environment develop moral attitudes, values, and behavior just as they grow and mature physically and mentally. According to this theory, just as it is both desirable and statistically normal that all humans become sexually mature and be able to understand abstract concepts, so it is desirable and normal that they develop an understanding of and commitment to what are believed by some psychologists to be certain universal moral principles. The theory is therefore an attempt to deal with both kinds of questions at once – that is, to answer questions about how we ought to act, and to tie this to theories of how we learn and hold values.

Moral Development

According to the theory of moral development, at about two years of age children begin to acquire a sense of right and wrong. They begin to understand that some of their friends are "nice" and "good" while others are not. Eventually, with age, an elementary understanding of fairness and justice develops and becomes an increasingly sophisticated outlook on problems of social interaction. Moral development is the growth from early childhood through adulthood of a person's abilities to

1. identify situations in which decisions and actions should be based on some concepts of justice, rightness, duty, and caring,
2. reason toward a choice of action in such a situation, and
3. follow through on that choice with appropriate action.

In 1932, Jean Piaget found that children progress through well-defined stages in their development of the concept of fairness, and that these stages are measurable.[2] Building on the work of Piaget, Lawrence Kohlberg

---

2.   J. Piaget, *The Moral Judgement of the Child*, trans. M. Gabian (The Free Press, New York, 1965; originally published in 1932).

extended this research to adolescents and young adults, using interviews every three or four years over a twenty-year longitudinal study.[3] Kohlberg presented each subject with hypothetical moral dilemmas and then, using a structured scoring system to evaluate the responses, he identified six stages of moral development. The criteria for these developmental stages are:

1. Each stage reveals qualitatively distinct modes of thinking, called "cognitive structures." These cognitive structures are the underlying organization of thought, and are not acquired skills or knowledge.
2. Individuals develop from lower to higher stages in an invariant sequence; environmental factors may influence the speed of the development but not the direction of it.
3. Stages are not skipped over. For example, people reasoning at Stage 3 are capable of reasoning at Stages 1 and 2 and, if they are in transition, at Stage 4, but not at Stage 5.
4. Development from one stage to another occurs when individuals encounter an experience that does not fit into their acquired cognitive structures. By assimilating the experience into their thinking, people develop new structures of thought, thus leading them into the next stage.

From his data, Kohlberg organized six Piagetian stages of moral development, in three levels. These are outlined below, based on summaries by Rest[4] and Colby.[5]

LEVEL 1   *Preconventional*

Stage 1: Obedience – we do what we are told in order to avoid punishment. At this stage, the child obeys the caretaker, and to be good is simply to do as told. The child sees no plan or purpose to the rules; they are simply to be obeyed.

Stage 2: Purposeful exchange – "we'll make a deal." The child realizes that doing what others want may result in *their* doing what the child wants. For example, "If I don't scream, I may get a lollipop."

---

3. L. Kohlberg, *Collected Papers on Moral Development and Moral Education* (Laboratory of Human Development, Harvard University, Cambridge, MA, 1973).
4. J. R. Rest, *Developments in Judging Moral Issues* (University of Minnesota Press, Minneapolis, 1979).
5. A. Colby, L. Kohlberg, J. Gibbs, and M. Lieberman, *Longitudinal Study of Moral Development*, Monograph of the Society for Research in Child Development, vol. 48 (1, serial 200) (University of Chicago Press, Chicago, 1983).

LEVEL 2   *Conventional*

Stage 3: Being a nice person – at this stage, the child understands that cooperation, not only on a deal-to-deal basis but as a style of relating to people, is beneficial. Stage 3 behavior often results in loyalty and trust in friendships.

Stage 4: Law and order – everyone should obey the law. If one simply follows the letter of the law, one is correct and "moral." Because laws apply to all, it is easy to anticipate the actions of fellow citizens. This social system is stable because everyone is expected to conform. Many adults make most if not all of their moral decisions at this stage.

LEVEL 3   *Postconventional or Principled*

Stage 5: Societal consensus – obey the laws of the majority. At this stage, the difference between a bad and a good law is recognized, and good laws are defined as those developed by the democratic process. As long as people's lives, liberties, and pursuit of happiness are guaranteed, and the laws are democratically developed as the laws of the majority, then to act morally is to follow these laws.

Stage 6: Recognition and acceptance of universal moral principles, such as equality of rights and dignity of all individuals. Most laws in democratic societies are based on such principles. If, however, a law does violate a universal principle, a person at Stage 6 will act according to the principle and not the law.

This is, of course, a theory, and there is some doubt that moral development is chronological or even biologically based. Even if we accept that there are just six distinct "levels," there's a fundamental inconsistency: How do we account for the enormous differences in moral development between and within societies? If the process is as rigid and universal as is claimed, then we wouldn't expect such differences. Consider physical development from birth through puberty to senescence and death – now *that's* universal and invariant, with rare exceptions. On the other hand, one could argue that moral development is culturally based, and that the morals developed by the children and young adults are those within that culture.

As we describe it, the theory in fact ignores differences in cultures, or rather it takes for granted the structure and values of modern Western culture. The concept of democracy can be traced to classical Greece, but both

Plato and Aristotle describe democracy only in order to deride it. The moral development theory as just outlined makes no sense in societies that do not have or want democratic political processes.

However, we could modify the theory. In Stage 5, we could say that "good laws are those developed by processes that take into account everyone's interests," just as a democratic process does. This would enable the theory to encompass nondemocratic societies in which a distinction is still made between legitimate and illegitimate sources and uses of power, but where majority vote is not the criterion of legitimacy. We could also modify Stage 6 and remove its reference to individual rights, since most cultures do not subscribe to the highly individualistic Western concept of rights.[6]

Alternatively, we could drop the unrealistic claims of the theory, which present moral development on a model similar to psychological development. In this case, the theory becomes simply an account of the development of individuals within Western-type societies where values such as autonomy, equality, and democracy are highly valued. This modification would make the theory more plausible, albeit less sweeping.

Professional Morals

Even with all of its shortcomings, we might suppose that the Kohlberg moral development concept is a valid approach to the evaluation of moral character. If so, we might be able to imagine that engineers can be classified according to similar moral categories, but not necessarily that the development occurs chronologically through an engineer's career. Indeed, younger engineers entering the profession might well operate at the highest moral levels, and the realities of life can cause a deterioration of these values. In the discussion that follows, we therefore categorize professional morals with no suggestion that these occur in chronological stages.

Professional engineers interact with society not only as professionals who fulfill the wishes of the society that has educated them, but also and perhaps foremost as human beings. The way in which moral values are translated

---

6.   Elsewhere, the junior author argues that this concept is both historically and logically tied to a par-
      ticular view of human nature and society that arose, uniquely, in sixteenth- and seventeenth-
      century Western Europe. A. S. Gunn, "Traditional Ethics and the Moral Status of Animals," *Envi-
      ronmental Ethics*, vol. 5, no. 2 (1983), pp. 133–153.

into professional practice determines how each individual behaves in professional situations where ethics may influence the selection of alternative courses of action. Professional engineers (as well as other professionals) by virtue of their training assume a responsibility to society and to the profession. Thus, professional engineers have, in addition to personal values, a second layer of values that apply to their professional conduct.

From this notion of the similitude between private moral development and professional ethics, Richard McCuen has suggested six categories of professional engineering morality. The following descriptions are derived from McCuen's ideas[7] but do not strictly adhere to his interpretations.

LEVEL 1   *Preprofessional*

Stage 1: At this level, the engineer is not concerned with social or professional responsibilities. Professional conduct is dictated by the gain for the individual, with no thought of how such conduct would affect the firm, the client – engineer relationship, or the profession.

Stage 2: Just as the child recognizes in Stage 2 that there is something to be gained by "being nice," the engineer at this stage connects conduct to marketability. Thus, while the engineer is aware of the ideas of loyalty to the firm, client confidence, and proper professional conduct, ethical behavior depends on the motive of self-advancement.

LEVEL 2   *Professional*

Stage 3: At this stage, the engineer puts loyalty to the firm above any other consideration. The firm dictates proper action, and the engineer is freed from further ethical considerations. Just as the child recognizes that cooperation is rewarded, so the engineer concentrates on technical matters, becomes a "team player" within the firm, and ignores the ramifications of the job on society and on the environment.

Stage 4: At this stage, the individual retains loyalty to the firm but recognizes that the firm is part of a larger profession, and that loyalty to the profession enhances the reputation of the firm and brings rewards to the engineer. Good engineering practice

7.   R. H. McCuen, "The Ethical Dimensions of Professionalism," *Journal of Professional Activities*, ASCE, vol. 105, no. E12 (April 1979).

becomes that which helps the profession – and not necessarily society in general. Many adults attain Stage 4 in their everyday moral conduct, where the laws of the land, regardless of how the laws were enacted, deserve their support. Engineers who follow the laws of professional conduct as dictated by the various professional societies act at Stage 4.

LEVEL 3   *Principled Professional*

Stage 5: Here service to human welfare is paramount. The engineer recognizes that such service will also bring credit to the firm and to the profession. Thus, the rules of society determine professional conduct. Where professional standards do not apply or are in conflict with the prevailing morals of society, society's values take precedence. As in Kohlberg's stages, the engineer at Stage 5 does not question the validity of the social rules, as long as the rules are arrived at by democratic social consensus.

Stage 6: At this stage, professional conduct follows rules of universal justice, fairness, and caring for fellow humans. This level is the most complex because acts of justice and caring can often contradict the prevailing social order and/or the professional code of ethical conduct. For example, until 1977 the Code of Ethics for the American Society of Civil Engineers had a footnote indicating, in effect, that although bribes were not condoned in most countries, where bribes were a standard method of doing business, bribery of public officials was an acceptable engineering practice. An engineer working at any stage other than Stage 6 would have thus used bribes as standard operating procedure whenever working in a country that tolerated bribery.

Awareness of professional morality helps us understand our behavior and the behavior of co-workers when faced with ethical decisions. Case studies like the ones in Chapter 1 can clarify these stages of professional engineering morality. Take, for example, the question of loyalty to a firm versus professional integrity, as illustrated in Case 6, "Just a Minor Editorial Change." Stan, an engineer, has developed a professional conclusion that the oil refinery will cause irreparable environmental damage and has written a report showing this. His clients have already seen the report, so it will be very difficult to retract it. Yet he has a direct order from his engineer boss, Bruce, either to withdraw or to discredit the report.

At what stage is Bruce acting? Obviously, he does not consider social welfare paramount and puts the firm and its business ahead of any social concerns. In this case he is acting at Stages 2 or 3. If Stan does as he is told, he is simply following orders and working at Stage 3. But Stan has other options. He could refuse to rewrite the report since this would require him to deceive the clients by suggesting a false conclusion. He would also be placed in a very bad light with his clients, who have already seen the draft copy. He could instead disengage the firm from the job and return the clients' payments, invoking (correctly) a conflict of interest. He would then put his loyalty to the profession in the forefront, and would be working at Stage 4. Stage 5 behavior would allow him to assign the job to someone else within the firm and thereby drop the whole matter. Or, if he felt strongly that the neighborhood group was speaking for the environment and that they badly needed his services, he could resign from his job and continue working with the environmental group, risking the animosity of his former colleagues in the firm. This would be Stage 6 behavior – putting universal principles above all other considerations.

We reiterate that although we recognize that engineers work as professionals at clearly defined moral stages, this does not mean there is a *development* of engineering morality throughout a professional career, paralleling Kohlberg's moral development. In fact, most young engineers tend to be highly idealistic and may enter the engineering practice in Kohlberg's Stage 5 or even 6, and with time and increased responsibility, begin to make decisions at lower stages. Thus, the idea of moral *development* in professional engineering is unlikely. In other words, contrary to Kohlberg's theories, an engineer may actually *regress* in professional ethical development. Nevertheless, the modes of ethical operation, based on Kohlberg's stages, may illuminate the various ways engineers act professionally.

2.3   As Engineers See Themselves

The American Society of Civil Engineers (ASCE) describes the engineering profession in this way:

> A profession is a calling in which special knowledge and skill are used in a distinctly intellectual plane in the service of mankind, and in which the successful expression of creative ability and application of professional knowledge are the primary rewards. There is implied the application of the highest standards of excellence in the education fields pre-

requisite to the calling, in the performance of services, and in the ethical conduct of its members.[8]

This is a rather cold and bland description. Engineering is, in fact, an exciting profession, and a very special one. Engineer and former United States president Herbert Hoover felt this way about engineering:

> It is a great profession. There is the fascination of watching a figment of the imagination emerge through the aid of science to a plan on paper. Then it moves to realization in stone or metal or energy. Then it brings jobs and homes to men. Then it elevates the standards of living and adds to the comforts of life. That is the engineer's high privilege. The great liability of the engineer compared to men of other professions is that his works are out in the open where all can see them. His acts, step by step, are in hard substance. He cannot bury his mistakes in the grave like the doctors. He cannot argue them into thin air or blame the judge like the lawyers. He cannot, like the architects, cover his failure with trees and vines. He cannot, like the politician, screen his shortcomings by blaming his opponents and hope that the people will forget. The engineer simply cannot deny that he did it. If his works do not work, he is damned forever.[9]

This special feeling leads to a degree of camaraderie that is seldom present in other professions. Even in college, engineering students form a strong bond among themselves. Unlike prelaw or premedical students, engineering undergraduates thrive in group activity, and unselfishly provide help to each other. The *problem* is what they are all attacking, not the professor or the grade curve. In Canada and many European countries, engineering graduates are inducted into a professional guild with an iron-ring ceremony, an occasion that results in lifelong bonding.

In the practice of engineering, there is also a feeling of freedom that other professions envy. Samuel Florman has eloquently described this as the "existential pleasures of engineering."[10] According to Florman, engineers have the freedom to concentrate on their job, unfettered by monetary and social concerns. Engineers do not make the decision to build a dam, for

---

8.   C. Nelson and S. R. Peterson, "If You're an Engineer, You're Probably a Utilitarian," *Issues in Engineering*, ASCE, vol. 108 no. EI1 (1982).

9.   Herbert Hoover, *The Memoirs of Herbert Hoover: Years of Adventure, 1874–1920* (MacMillan, New York, 1951), pp. 132–133.

10. Samuel Florman, *The Existential Pleasures of Engineering* (St. Martin's Press, New York, 1976).

example, but are simply asked to design and construct it, thus freeing them to perform the technical task for which they were trained and which they find most pleasurable. Florman argues that contrary to the popular notion of engineering being highly restrictive, engineers in fact have the highest level of freedom of all the professions.

Engineering is also an honored profession. Some of the earliest engineers in the newly founded United States were our best minds and leaders. George Washington taught himself surveying and supervised the construction of roads, canals, and locks. Thomas Jefferson was an inveterate tinkerer and some of his gadgets are clever even by modern mechanical engineering standards.

Engineers are indeed a special breed. A study done some years ago evaluated the characteristics of students who entered college intending to study engineering. The study found that those students who continued in the engineering program after two years differed markedly from those who left engineering.[11] Only certain persons choose to become engineers, and to undertake the rigors of the demanding college curriculum that allows entrance to the field. Engineering students and practicing engineers tend to have a very positive image of themselves and a lively *esprit de corps*. Engineering is *fun*.

Engineers also see themselves as performing a public service. The ASCE motto describing the civil engineering profession as the "people-serving profession" neatly sums up engineers' perception of themselves. Engineers build civilizations. Engineers serve the public's needs.

2.4   As the Public Sees Engineers

While engineers see themselves as successful problem-solvers acting in the public interest, the public's perception of them is often somewhat different. Well-publicized engineering failures resulting in damage to human and environmental health have encouraged many people to perceive engineers as the creators, not solvers, of problems.

This view is not restricted to North America. A recent poll carried out in New Zealand and Australia asked people to rate various occupational groups for ethics and honesty. In New Zealand, only 44 percent rated engineers as "high" or "very high," admittedly well ahead of accountants, lawyers, real

---

11.   C. F. Elton and H. A. Rose, "Students Who Leave Engineering," *Environmental Education*, vol. 62, no. 1 (1971), pp. 30–32.

estate agents, and business managers, but well behind pharmacists, police, doctors, schoolteachers, and dentists. Australians have a higher opinion of engineers, 56 percent rating them "high" or "very high" but still behind health professionals and teachers.[12] Of course, polls themselves can be unreliable. In a classic paradox, this poll also noted that only 11 percent of New Zealanders and 8 percent of Australians gave a high rating to journalists. That aside, engineers should be seriously concerned that about half of the people questioned in a scientifically conducted poll think that the profession's ethical standards are less than high.

Part of the public attitude toward engineers might be explained by the engineers' role as one of conducting social experiments, as pointed out by Martin and Schinzinger.[13] Engineers do not have all the answers at hand when a problem arises and often cannot perform full-scale experiments to obtain such answers. For example, when designing a suspension bridge, engineers cannot construct a full-scale model of the bridge to test the design. Instead, they use the best knowledge and designs available and extrapolate, using sound judgment. Sometimes the extrapolation is faulty and the experiment fails. Engineers learn from such failures and modify the design process in subsequent projects.[14] What is important is that the experiments occur not in a private laboratory but on the public stage.

Engineering experiments affect humans in many ways. A road, for example, is both a technical and a social undertaking. The road might affect homes and businesses, create traffic noise, or even divert funds from other projects. A new lawn mower is designed and manufactured and it might cut grass effectively, but at the expense of neighborhood quiet and perhaps even auditory damage to the user. A reservoir is constructed, but at a cost to recreational whitewater rafting, or despoliation of a bottomland virgin ecosystem, or the destruction of an ancestral homestead. A new biological weapon, designed to kill people, is developed with the knowledge that its use would devastate the global ecosystem. How is the engineer to deal with these conflicting public benefits and costs?

The fundamental difference between how engineers see themselves and how the public sees them depends on how well engineers meet public concerns. Sometimes engineers are labeled the "tools of the establishment," or the "despoilers of the environment" or the "diligent destroyers." With the

12.   *Time*, 31 May 1993.
13.   Mike W. Martin and Roland Schinzinger, *Ethics in Engineering* (McGraw-Hill, New York, 1989).
14.   Henry Petroski, *To Engineer Is Human* (St. Martin's Press, New York, 1985).

public, it's often "we" against "them," and "they" are too often the engineers. How come the engineers are the villains? Why do the engineers so often appear to be the bad guys or, as in the old cowboy movies, the guys wearing the black hats?

The source of the problem is that engineers and the lay public hold different ethical attitudes. Because engineers tend to be utilitarians, they look at the overall and aggregate net benefit, thus diminishing the importance of harm to the individual. Because engineers are positivists, they tend to ignore or dismiss considerations for which reasons of a certain type cannot be given – for example, quantifiable or at least empirical data – thus ignoring intangibles. Of course engineers value people, but they have a particular view of what is good for people and how this good is to be determined in a given case. Finally, engineers think of themselves as doing applied physical science, not applied social science. The physical science approach to engineering (ignoring the "people-serving profession" motto) allows engineers to think of their work as not being germane to the needs of society. This conflict of ethical outlooks is a root cause of much of the problem with engineers' interaction with the public.

Engineers as Utilitarians

Utilitarianism, as we explain in Chapter 4, is an ethical theory. It holds that the right action is the one resulting in the greatest utility, that is, the greatest net benefit to all concerned. In traditional utilitarian theory, benefit is measured in terms of happiness or pleasure, although today it is usually explained in terms of interests or preference satisfaction. Cost-benefit analysis is a fairly straightforward application of utilitarianism in that both the sum of the costs and the sum of the benefits can be calculated and one divided by the other.

Critics of utilitarianism argue that utilitarians will accept some suffering as long as the net effect is positive. For example, utilitarians would accept the sacrifice of a random human life in order to achieve some net good that cannot be achieved without such a cost. The utilitarian value system thus allows the U.S. Environmental Protection Agency (EPA) to calculate the risk associated with pollutants and to place an "acceptable" risk at a probability of 1 in 1×10^6. In utilitarianism the result of one's actions is what matters, not the quality of the action itself. The harm of one person in a million is a negligible cost if the benefits are great. Utilitarian calculus might even argue for the human sacrifice in Case 7 if the benefits of the power station will be greater for a large number of people.

Most people, however, are not utilitarians when it comes to actions concerning their own welfare or even the welfare of others. Even if people are not by nature selfish egoists, it is certainly expecting a lot of people to weigh their own interests and those of their loved ones equally and dispassionately against the interests of others. But even if we can be sufficiently dispassionate, or if our own interests are not directly involved, cost-benefit analysis may be considered unethical since the costs and benefits can be (and most likely are) disproportionately distributed. Some people will suffer more, others will gain more, and many will consider the resulting distribution unfair. Critics of utilitarianism argue that if you would not accept a situation where the one person being harmed is yourself, or someone you love, then it must be unfair to accept that some other one person in a million will be harmed, since for that one person the imposition of harm is also unfair.

Many people also believe it is unethical to place a value on a human life, and refuse even to discuss the one-in-a-million death caused by a given environmental contaminant. They do not accept the validity of the cost-benefit calculations when human lives are involved. They thus reject common calculations for highway improvements that assign monetary values to human lives.[15] Many people hold that such arithmetic is not possible because they believe human life is of infinite value, or that whatever its value is, it cannot be measured in dollar terms. But if there is no measure of lives in dollars, comparing the costs of highway improvements and of highway deaths is like trying to divide apples by oranges. And, of course, if human lives are infinitely valuable, they cannot possibly be figured into an equation.

Many engineering consultants, therefore, soon discover that utilitarian calculus, so valued as a decision-making tool by engineers, is largely irrelevant to most individuals. In a public hearing, an engineer announces that the net detrimental effect of the emission from a proposed fossil fuel power plant will increase cancer deaths by only 1 in 1×10^6, and considers this risk to be quite acceptable given the benefit of the power produced. But many members of the public, rejecting utilitarian calculus as a basis for decision making, will consider this one death in a million as grossly unethical. Frus-

---

15. For example, if a certain improvement will have an annualized cost of $500,000, and if this improvement will save an estimated four lives per year, and if a human life is valued at $200,000, then it makes sense to proceed with the construction. If, however, only two lives would be saved by the improvement, then $2 \times \$200,000$ is less than $500,000 and the construction is not warranted.

trated engineers often dismiss such lay resistance as "technical illiteracy." But this is a mistake. It is not technical illiteracy but a different ethical viewpoint that is in question.

Of course, cost-benefit analysis is often inevitable. Without it, necessary decisions – such as where to place a landfill or whether to use hazardous materials in consumer products – remain unresolved. Opponents of cost-benefit analysis themselves often resort to some form of it, but only after exhausting all other alternatives.

Engineers as Positivists

The second problem engineers have in interacting with society is that unlike most people, engineers in their work are inevitably positivists. Positivism is an attempt to apply the empiricist tradition of Francis Bacon and Isaac Newton and others to all elements of life.[16] Positivists believe that all sciences are going through an inevitable evolution from the spiritual to the positivist position, where all knowledge is based on the scientific method. In terms of positivism, mathematics is clearly the most developed of the sciences, and engineering, which is firmly grounded in mathematics, is perhaps second.

Positivists believe that objects of study in the social sciences should be treated in the same manner as objects of study in the physical sciences; that is, the observer should and can remain apart and separate from the object under study. They believe that the "physical sciences provide the paradigm of objective knowledge."[17] Science is value-neutral, and ethical concerns are only expressions of emotions. Since true science is objective, and since ethics cannot be "done" using the scientific method, ethical considerations have no bearing on scientific practice. Physical sciences are indeed largely based on positivist ideas and, by extension, so is engineering.

Positivists also assert the value of the scientific method for social science. They argue that social investigations can also be conducted in a value-free environment, and that social scientists should eliminate all bias and emotional involvement from their investigations. Positivists hold that if there is disagreement between any two rational people, all one needs to resolve the conflict are more data and better experiments.

16. J. K. Smith, "Quantitative Versus Qualitative Research: An Attempt to Clarify the Issue," *Educational Researcher* (March 1983).
17. R. Baum, "A Philosophical/Historical Perspective on Contemporary Concerns and Trends in the Area of Science and Values," *Science, Technology and Human Values*, vol. 2, no. 1 (1974).

This view is, of course, not shared by all, and there are numerous alternative approaches. Some social scientists challenge the attempt to apply positivism to the social sciences on the grounds that social sciences are concerned not with *things* but, rather, with the notoriously subjective human mind. In their view, the observer is inevitably an integral part of the study since his/her presence affects the outcome. This so-called *idealist* school[18] does not accept the possibility of developing a set of universal laws to explain and predict human conduct, since the very act of obtaining the data is dependent on who gathers the data. For idealists, social science must therefore be descriptive instead of explanatory or predictive.

In idealist theory, scientists can resolve disagreement only by appealing for understanding, since new data are unlikely to be helpful. Data are useful only within the context of how they were obtained. The situation that occurred when the data were taken can never occur again, and whatever data are subsequently obtained cannot resolve the issue. As Taylor observes, "The superiority of one position over another will thus consist of this, that from the more adequate position one can understand one's own stand and that of one's opponents, but not the other way around."[19] That is, if your arguments allow you to understand and appreciate the opposite view, and this is not possible with the opposing argument, then your view is the superior one.

In the idealist school, facts are inseparable from values, and every scientific experiment is value-laden. First the scientist decides what kind of experiment of all those that could be performed he/she actually chooses to undertake, how to conduct it, and what information to obtain. Second, truthfulness is presumed, and scientists must be able to trust the results of previous studies. Third, scientists choose what results to publish and what conclusions to draw based on their work. Fourth, the agencies that fund research have their own values and display them in the way they provide money to the researchers. Finally, whether or not the findings actually get to be published, and if so, in what form, depends on their acceptability to the scientific establishment. The scientific establishment, which controls conferences, journals, and other means of disseminating ideas, determines whether the findings are worthy of publication.

As idealists point out, the axioms underpinning positivism cannot be

18. J. K. Smith, "Quantitative Versus Qualitative Research," pp. 3–13.
19. C. Taylor, "Interpretation in the Science of Man" (1977), quoted in J. Wilson, *Social Theory* (Prentice-Hall, Englewood Cliffs, NJ, 1983).

independently proven and positivists must paradoxically accept these on faith. Idealists ask how the positivist position differs from the message of religious missionaries, who, being absolutely sure of their own faith, populate our television screens or travel to distant parts of the world to indoctrinate the masses. Do positivists, like missionaries, indoctrinate the public, but more successfully?

Unfortunately, the education and training of modern engineers is usually too positivist for their own good. Engineers are carefully taught to be pragmatic, logical, rational, sensible, and systematic in their approach to problems. These are useful and absolutely necessary attributes, but they also tend to alienate the engineer from the rest of society.

Many engineering decisions are clearly value-laden, and positivist thinking helps very little in resolving such dilemmas. Since engineers have been taught to uphold positivism as a truth, they are frustrated when their logical approach does not result in an acceptable resolution to value-laden problems. Positivism cannot help engineers respond to the question "All things considered, what *ought* I to do?"

Engineers as Applied Physical Scientists

Engineering as an academic discipline has always been grouped with the physical sciences, since engineering is supposed to deal with *things*, not people. Physical sciences such as chemistry deal with atoms and molecules – things. Chemistry is a science that describes our world in chemical terms, and the propositions of chemistry would be true even in the absence of people. Water would still be two atoms of hydrogen and one of oxygen, whether people knew or cared about it. Chemists labor in the laboratory to discover new knowledge, which they believe to be timelessly true. Moreover, they typically see themselves as seeking this knowledge not because they hope to benefit humanity, but because they believe that an advance in knowledge is good for its own sake. Some scientists and mathematicians have gone out of their way to emphasize the "purity" of their work. G. H. Hardy, the eminent Cambridge University mathematician, boasted: "I have never done anything 'useful'. No discovery of mine has made, or is likely to make, directly or indirectly, for good or ill, the least difference to the amenity of the world."[20] Presumably Hardy's colleagues shared his attitude, for the toast at the Cambridge Mathematics

20. Quoted by John Fulton in D. Daiches, ed., *The Idea of a New University* (Deutsch, London, 1964). Contemptuous quotation marks in the original.

Christmas party was for many years, "To pure mathematics: may she never be of use to anyone!" While the accuracy of this image of science may be questionable, it is nonetheless how many scientists see themselves.

By contrast, social sciences focus on *people*, not things. Engineering is also directly concerned with people. In fact, *engineering without people cannot exist*. Engineering is the application of the sciences *for the benefit of people*, otherwise there would be no reason for engineering. Engineering takes the knowledge created by the sciences and applies it for the benefit of the people in order to create a higher standard of health, comfort, and living. This is also the view of at least some scientists. As Roger Guillemin writes, "The use, including the misuse or ill-use of . . . knowledge is the realm of politicians, engineers and technologists."[21]

Because engineering is inextricably tied to people, it cannot be just an applied physical science; it is also an applied *social* science.

This said, one reason why engineers often have difficulties in their interactions with the public is that they misinterpret their own profession. They believe they can safely hide their work from public scrutiny and happily work with their machines, buildings, and refineries. But engineers cannot be thus insulated, since their entire raison d'être is to serve the public. It is therefore no wonder that the public will not leave engineers in peace.

In summary, engineers (for better of worse) are utilitarians and positivists, and generally see themselves as applied physical scientists who make value-free decisions. As Nelson and Peterson have pointed out, one reason engineers are utilitarians might be precisely "because they cannot be held morally accountable afterwards."[22] In the engineers' view, they function as trustees for the greater public good.

The public, however, does not see knowledge in positivist terms and does not employ utilitarian calculus in making value-laden decisions. While there is no one philosophy or ethical theory to which the public in general subscribes, we suggest that a concern for the interests of oneself and family and friends is of much more significance for most people than a dispassionate calculation of the costs and benefits to all concerned. This is what makes it so difficult for engineers to understand their own function in society, and for the public to appreciate the role of the engineer.

21. Quoted in S. A. Lakoff, "Moral Responsibility and the 'Galilean Imperative'," *Ethics,* vol. 91 (1980).
22. C. Nelson and S. R. Peterson, "A Moral Appraisal of Cost-Benefit Analysis," *Issues in Engineering,* ASCE, vol. 108, no. EI1 (1982).

One solution to this problem is to educate the public to be more positivist and utilitarian in its outlook. In the 1920s and 1930s, the technocracy movement attempted just this, and the United States even elected its only engineering president, Herbert Hoover. But success has been limited. The best efforts of the schools have not advanced the cause of positivism among the general public. Could it be that the public simply does not want to be converted? Perhaps it does not *want* to believe in the positivist "religion"?

If this is the case, engineers ought to recognize and appreciate this different outlook. At times, engineers should practice their profession as applied *social* scientists instead of applied *physical* scientists. By so doing, they would operate according to values familiar to the general public and thus would enhance their worth to society.

2.5 The Engineer's Responsibilities to Economically Deprived Peoples, to Distant Peoples, and to Future Peoples

Engineering as a profession cannot exist without the public. Engineers are professionally responsible to the public, and the engineering codes of ethics all refer to the "paramount" obligation of engineers to the welfare of the public.

But what exactly is the public? Certainly the public includes the clients, employers, and users of the products of engineering. But what about those people who are distant economically, spatially, and temporally from the engineer's actions?

Deprived Peoples

During the past few years a new term has gained importance to environmental engineers – "environmental racism."[23] This idea originates from the empirical evidence that undesirable land uses such as incinerators, wastewater treatment plants, landfills, and the like are often sited in areas of a community with a high percentage of minorities. The undesirable land use therefore is unevenly and unjustly distributed, and this is interpreted as being racially motivated.

---

23. Robert D. Bullard, "Environmental Blackmail in Minority Communities," in B. Bryant and P. Mohai, eds., *Race and the Incidence of Environmental Hazard* (Westview Press, Boulder, 1992), pp. 82–94.

Although we do not want to support the uneven distribution of undesirable development, we believe that the uneven distribution of undesirable land use is much more likely due to economic factors – the land is simply less expensive in the poorer parts of town, and these areas are often the minority neighborhoods. Engineers and other decision makers are not necessarily racists, and they *are* beholden to the public to provide services at the least cost.

In the past the uneven distribution of social costs has not been factored into the decision making. For example, the community landfill in Chapel Hill, North Carolina, was sited near a minority neighborhood because the land was inexpensive and the level of public opposition at the time of the siting process was low. The neighbors were assured that the landfill would last for only about ten years and then a new landfill would be constructed in another part of town. When ten years had passed and a new landfill became necessary, the town decided to enlarge the existing landfill because the problem of siting new landfills became politically too difficult. The town officials told the people in the minority neighborhood that the promises to move the landfill to another location could not be binding because these promises had been made by a previous administration.

There is clear injustice here, and breaking promises made by previous local administrations is blatantly immoral. Is this, however, a case of racism? Or is it a case of gutless politicians simply breaking promises?

Whatever the reason for injustices in the imposition of societal cost such as undesirable land use, the concept of "environmental racism" has added another variable into the decision making of engineers and governmental agencies. There is no ethical justification for the unequal treatment of citizens, regardless of race or economic standing. The recognition of such inequities has made it quite clear that economics will no longer be the sole criterion governing the location of undesirable land use and the distribution of societal costs: and this is how it should be.

Distant Peoples

Most of the literature on environmental ethics pays little attention to the well-being of people in both poor and distant countries. This criticism is not new. As British ecologist John Black wrote in 1970:

> We should remind ourselves that we are viewing the world from a position of privilege, that the problems of western civilization are not nec-

essarily those of the rest of the world, and that it is one thing to complain about pollution and loss of scenic amenity on a full stomach, quite
another to be faced with the ever-present risk of starvation.[24]

Although Black's book was widely read, few people seem to have taken his
point. The problem is not that people do not care about others, but that they
too often discuss nature without reference to humans. In the United States,
for example, environmentalists urge nations in the developing world to preserve their rain forests and reduce chlorofluorocarbon (CFC) and carbon
dioxide emissions in order to prevent global warming, protect the ozone
layer, and preserve the "global heritage" of biodiversity. But many people in
developing nations consider such arguments to be little more than neocolo-
nialist attempts to promote the selfish interests of rich nations. Besides, it
was the colonial powers that began and still perpetuate the world's environmental degradation. In the words of Malaysian prime minister Dr. Mahathir
Mohamad,

> All sorts of campaigns are mounted by the richer countries against
> [destroying forests in order to generate] hydroelectricity, which have
> already developed their hydroelectric potential. Now of course the
> World Bank will be used to deprive poor countries of cheap hydroelec-
> tric power. Can we be blamed if we think that this is a plot to keep us
> poor?[25]

The prime minister has a valid point; the rhetoric of global heritage, in
the absence of a concern for our fellow humans, is quite inappropriate. A
shared heritage requires a single community. A global heritage therefore
assumes a global community in which all people share resources.

But this is not how our world operates. Rich countries have severely damaged both their own environments and those of most poor countries, either
through direct colonial rule or through neocolonialism, trade, and resource
exploitation. If environmentalists in the wealthy nations are to have some
influence on environmental policy in the sovereign poorer nations, they
must begin with their own governments. In particular, they must change
their own states' policies of expropriating most of the world's resources and
effectively excluding poorer nations from decisions about "common"

---

24. J. Black, *The Dominion of Man: The Search for Ecological Responsibility* (Edinburgh University Press,
     Edinburgh, 1970), p. 139.
25. Mahathir Mohamad, quoted in *New Straits Times*, 26 September 1991.

resources, in particular the oceans and Antarctica. Any environmental ethic should take account of the needs of all humans.

Future Peoples

In the United States and many other countries, environmental decision-making has become increasingly public and consultative. A decision to construct a power station, a prison, or a waste incinerator affects many people. These people rightly have a say in the planning process through formal and informal participation including consultation, lobbying, and public hearings. The opportunity for class action suits and the relative ease with which United States law grants standing to concerned individuals in environmental cases provides additional access for public involvement. Despite all this representation, one large group of people receives no hearing. These are the people yet to be born.

The effects of today's management decisions will be felt for years to come. Indeed, many decisions have no effects until decades later. For example, it may take generations for waste containers to corrode, for their contents to leach, for the leachate to migrate and pollute groundwater, and for toxic effects to occur. The only persons to be affected adversely by a management decision may be the only persons who have had no say in the decision. They cannot speak up for themselves and, unlike children, they have no one legally responsible for representing their interests. In fact, we often deliberately ignore the interests of future generations. For decades we have produced, and continue to produce, long-lived radioactive wastes, without any agreement about how to keep them safe in the future or even whether they can be kept safe at all. If we were serious about protecting future generations, we would not generate the waste until we have figured out how to manage it safely.

Instead, there is no clear long-term responsibility for hazardous waste management. United States legislation is much more comprehensive and in principle tougher than most. The main statute, the Resource Conservation and Recovery Act of 1976, requires waste facility operators to provide "perpetual care" for depositories. "Perpetual care," however, is defined as a period of only thirty years. In terms of the interests of future generations, when we say "perpetual" we should *mean* perpetual, and not just a few decades.

We cannot, of course, be sure that future generations will actually suffer harm if we mismanage our rsources, but if it is wrong to cause harm, it is also wrong to impose risks of harm. All currently proposed methods of dealing with high-level radioactive waste, including temporary storage until a method is agreed upon, impose serious risks on future as well as present generations. According to the arguments presented so far, it is unethical to impose those risks on future generations.

Many Americans were outraged at proposals to export hazardous wastes to poorer nations in the late 1970s and early 1980s,[26] resulting in legislation limiting (though not banning) the practice. Surely if we oppose sending hazardous materials into other parts of the world, we should also oppose attempts to export them into the future. In this view, our current practices and policies represent a form of discrimination. Just as we condemn racism and sexism, so we might condemn "presentism": the systematic ignoring or setting aside of the interests of people who have yet to be born.

Philosophers and others have raised various objections to the claim that we ought to forgo benefits (or incur costs) today in order to provide benefits (or reduce costs) for future generations. Defenders of presentism make several interesting points, but none of them convincingly challenges our ethical obligations to future people.

The first argument advanced by presentists is that since future generations do not exist, they cannot be the subject of obligations. Even to talk of "them" is misleading, as it suggests a crowd of as yet unconceived beings floating in a world of preexistence.[27] There are no nonexistent beings, and therefore we cannot have any obligations to future generations.

A second objection is that we cannot predict the future. We have no idea what future people will want or value, and so we cannot take their interests into account. Imagine what our ancestors might have supposed we today would need most: horseshoes, or spokeshaves, perhaps a plentiful supply of leeches, Epsom salts, powdered unicorn horn and tincture of opium, knowledge of alchemy, the Ptolemaic system or phlogiston theory? Moreover, we cannot assume that future people will share our preferences, inclinations, or values. Our ancestors baited bears, stoned adulterers, denied women the vote, and promoted slavery, witchburning, torture, and crusades against the heathen. Perhaps we even bear an obligation *not* to plan for future people. However well-intentioned we may be, we are likely to get it wrong.[28]

26. H. Shue, "Exporting Hazards," *Ethics,* vol. 91 (1981), pp. 579–606.
27. P. Singer, *Animal Liberation* (Avon Books, New York, 1975).
28. M. Golding, "Obligations to Future Generations," *Monist,* vol. 56 (1972), pp. 85–99.

Third, obligation is not just a matter of identifying others' wants, needs, and interests. You may want and need my car, but that does not oblige me to give you my car. Some philosophers and political theorists argue that obligations exist only among members of a community linked by reciprocal rights and duties.[29] You can acquire a right to my car only if we are both willing to enter into an agreement, such as a purchase contract. More generally, in a community we all have obligations to respect each other's rights, based on a recognition by others of our own rights. But we do not form any kind of community with the future. We can do things that may turn out to affect them (though we do now know what the effects, if any, will be), but future people can do nothing to affect us. As Joseph Addison observed in 1714, " 'We are always doing something for Posterity,' says he 'but I fain to see Posterity do something for us.' "[30] In this argument, while we might act nobly when we take account of future generations, we bear no ethical obligations to do so.

Finally, some presentists argue that we have no obligation to sacrifice our interests for the sake of future people because they will be better off than we are. Indeed, the farther in the future the people are, the better off will they be. One way of analyzing this is to "discount the future"; to assume that a resource we value today will be worth less in the future. Thus, it makes some sense to extract the resource now when its value is greatest. Most economists accept this argument as valid and only debate the rate at which future interests are to be discounted. It therefore makes little sense to save resources for future generations.[31]

Alternatively, presentists argue that each generation adds to the total stock of wealth, of knowledge, art, technology, structures, and everything else that enriches our lives. As empirical evidence shows, each generation is richer than those before it, but poorer than those that follow after. Thus, "Future people may inherit fewer resources than we have, but will be compensated for this by inheriting improved technology and accumulated capital."[32] We cannot do anything about the fact that we have more than our

29. For example, J. Passmore, *Man's Responsibilities for Nature* (Duckworth, London 1974); or R. A. Watson, "Self-Consciousness and the Rights of Non-Human Animals in Nature," *Environmental Ethics*, vol. 1 (1979), pp. 99–129.
30. Joseph Addison, *The Spectator*, no. 583 (20 August 1714), quoted in *The Oxford Dictionary of Quotations* (4th ed.), ed. Angela Partington (Oxford University Press, New York, 1992), p. 4.
31. The assumption that inflation is a permanent economic condition is a common misrepresentation. Since the 1890s, the global economy has been largely inflationary, but before then, *deflation* was common. There is no guarantee that deflation might not begin again.
32. D. M. Hunt, "Responsibilities to Future People," in J. Howell, ed., *Environment and Ethics – a New Zealand Contribution* (Wellington, New Zealand Environmental Council, 1986), pp. 51–75.

great-grandparents had. Conversely, as the poorest of the poor, we surely have no obligation to make ourselves even poorer in relation to the future. In deciding how to manage hazardous waste we should certainly consider our fellow citizens, but not the future. Instead, we can remain confident that whatever we do, future generations will be better off than we are. In fact, we may even have a duty to help the deprived members of this generation, at the expense of future generations, in order to reduce the deprivation of today relative to tomorrow's abundance.

A related argument is that the history of technology is a history of progress. Dark Ages excepted, it has been a long time since any worthwhile technology was forgotten. As you sit at your desk in your air-conditioned room with a cup of freshly made coffee, listening to Beethoven on a portable compact disc player, you cannot deny that you are much better off than your subsistence-farming ancestors. Considering the enormous progress in medicine, transport, communications, and information processing in recent decades, shouldn't we expect that future generations will take care of any problems we bequeath to them?

With all these reasons for presentism, perhaps there is nothing wrong with our generating nuclear wastes in order to enjoy the benefits of electricity, or interpreting our obligations of "perpetual care" as extending for only thirty years. We can then conclude that we should avoid causing harm to others, but not to future generations, because:

- Future generations do not exist.
- We cannot know what they will need or want.
- The alleged obligation has no basis because we do not form a community with them.
- They will in any case be better off than we are, so they do not need us to look out for their interests.

These arguments for presentism, however, all contain serious flaws.

First, even if future generations do not yet exist (by definition), we can still have obligations to them. Certainly we do not bear the kinds of one-on-one obligations that contemporaries have to each other. For instance, ordinary debts and promises can be owed only by one person to another. But we also hold common obligations to anyone who might be harmed by our actions even though we do not know her or his identity. Indeed, we ought to avoid actions that could harm others, even when, given the circumstances, no one would have been harmed. Thus, we stop at red lights even if it appears to be safe enough to cross without stopping.

As an analogy, consider a terrorist who plants a bomb in a primary school. Plainly the act is wrong, and in breach of a general obligation not to cause (or recklessly risk) harm to fellow citizens. This is so even though the terrorist does not know who the children are, or even whether any children will be harmed (the bomb may not explode, or it may be discovered in time, or the school may be closed down). It is equally wrong to place a time bomb, set to go off in the school in twenty years time, even though any children who may be in the school when it goes off are not yet born. An unsafe hazardous waste dump is analogous to a time bomb. Indeed, the time bomb metaphor is so convincing that it has become hackneyed in popular accounts of waste dumps and other chemical hazards.

The analogy is not perfect, because the terrorist actively desires the bomb to kill the children, whereas waste facility designers and owners have no such motive. But is there much difference, ethically, between the person who fires a nuclear missile at random, knowing that it may kill someone, and one who deliberately targets a city? Anyone who risks causing serious harm to others, without excuse, acts wrongly, even if the identity of those at risk is unknown. We can be reasonably sure that there will be some people around for as long as our hazardous wastes remain hazardous, unless this or some future generation renders the earth uninhabitable for humans. It is not as if the people who will be around in the future will be imaginary or fictional. Dick Tracy and Johnny Appleseed will never be real, but twenty-second-century Americans will be as real in their time as twentieth-century Americans are today.

The response to the second claim, that we cannot know what future generations will want or value, is that nothing follows from it. That we do not know something is an argument for further study, not for adopting one course of action over another. Even if we have no way of knowing whether our descendants will put positive or negative values on cancer, chromosome damage, and birth defects, we are still not entitled to go ahead and assume they will not care about environmental health. If we have no basis for predictions, they are equally likely to be more concerned than we are. Certainly, there are changes, fashions, and revolutions in values and scientific theories, but there is stability as well. Future people will not be just like us. They may not enjoy Fosters, McDonald's, *Saturday Night Live*, trips to Disney World, baseball, or political jokes. More seriously, they may have quite different views on health care, abortion, euthanasia, preservation of old-growth forests, and defense spending. Despite the likely changes, we can assume that many values and priorities will remain stable. As Brian Barry writes:

> Of course we don't know what the precise tastes of our remote descendants will be, but they are unlikely to include a desire for skin cancer, soil erosion, or the inundation of low-lying areas as a result of the melting of the icecaps.[33]

Nor are future generations likely to want to suffer genetic damage or to produce babies with severe birth defects. Therefore, we can assume they will probably not want to be exposed to wastes that could produce these conditions.

Third, accepting that there will likely be people around in the future who share our lack of enthusiasm for skin cancer and soil erosion, do we have any obligations to them? True, we do not form a spatiotemporal community with them. But as the Scottish philosopher David Hume noted more than two hundred years ago, we are not indifferent to people in distant lands and other times, even though we shall never see them.[34] The continuing popularity of science fiction and fantasy literature demonstrates that our interest in the future is as great as our interest in the past. Nor is it true that we feel obligations only to those with whom we have reciprocal, quasi-contractual relations. We have obligations within our society to infants, the irreversibly senile, and people in comas even though we cannot communicate with them and have no expectations of receiving anything in return.

Even if we do not accept an obligation to provide benefits for nonreciprocating members of a community, we would all agree it is wrong to cause harm to them – that the wrongfulness of causing harm is independent of reciprocity or spatiotemporal community. If so, we ought not to harm or place at risk future generations either.

The fourth presentist argument is that future generations will be better off than we are, so we need not worry about their well-being. A first answer to this argument is that we still hold an obligation not to cause harm even though the people appear to be better off than we are. Mere disparity of wealth does not entitle poor people to steal from rich people, even if their poverty is the the result of undeserved misfortune. Furthermore, why

33. B. Barry, "Justice Between Generations," in P. M. Hacker and J. Raz, eds., *Law, Morality and Society* (Clarendon Press, Oxford, 1979).
34. D. Hume, *An Enquiry Concerning the Principles of Morals*, vol. 1 (1777).

should future generations bear any of the costs of our use of technology while we benefit from that technology? Future people might have no need at all for aerosol sprays, mercury batteries, and throwaway cameras; and they may not think the benefits of nuclear-derived electricity are worth the risks.

Presentism requires us to assume that technological solutions exist for what appear to be intractable problems. But perhaps there is no safe way to dispose of high-level radioactive waste; or if there were, it would require a degree of social, economic, and political stability that might not exist in the future.[35] What if our era of rapid technological progress comes to an end, at least for a time? Technology has provided a high level of material security and comfort for hundreds of millions of people over the last several hundred years. But at the same time there are also more sick and starving people in the world than ever before. Optimism about technological progress also assumes that nonrenewable material and energy resources are indefinitely substitutable, but there is no basis for this assumption. More subtly (and more speculatively) some scientists argue that human demands on the biosphere have reduced ecosystem resilience, so that what appear to be sustainable yields cannot last.[36]

The optimistic assumptions discussed earlier may turn out to be true, but we are not entitled to assume that they will be. In case they are not, we have an obligation to avoid creating detrimental environmental conditions for future generations that we would not create for ourselves. In terms of waste management, this means it is necessary to develop methods for safe storage over a very long time and, more important, to reduce steadily the quantity, toxicity, and persistence of the wastes we produce so that acceptably safe management becomes possible. Further, because we can assume that future generations will appreciate clean air and water, adequate wilderness areas, and the preservation of natural monuments, we do bear an obligation to the future.

2.6   The Engineer's Responsibilities to the Environment

Every person has predispositions or moral beliefs concerning non-human nature and its worth relative to the needs of humans. Carole in Case 10, for

---

35. Editorial, "Identification of Nuclear Waste Sites over Ten Millennia," *Nuclear and Chemical Waste Management,* vol. 6 (1988), pp. 95–100.
36. B. S. Norton, "Conservation and Preservation: A Conceptual Rehabilitation," *Environmental Ethics,* vol. 8 (1986), pp. 195–220.

example, has a basic moral value that killing innocent animals for no good reason is morally unacceptable. But she could not use this moral value to construct an ethical framework that would justify her actions to others. In instinctively avoiding the turtle on the road, Carole acted on moral belief, not ethical analysis. So how exactly can we develop a system of ethics for the environment?

The logical starting point in the search for environmental ethics might be the classical ethical theories. Ethics has attracted the attention of intelligent people for thousands of years. In all that time the sages must have had something useful to say that will be of practical use in the resolution of conflicts involving people and nature.

Surprisingly, however, the ethicists disappoint us. In the West, at least, ethics is, by definition and tradition, the discipline that tries to resolve questions of how people ought to treat *other people*. Nature matters only as a set of resources, and only to the extent that it contributes to the interests and welfare of human beings. The central concepts in conventional ethics, such as rights and justice, do not work very well when applied to animals, let alone plants, places, and natural objects.

Some ethicists, known as moral extensionists, have tried to broaden conventional ethical theories to take account of non-human nature and natural objects. As discussed in Chapter 4, such attempts to apply classical ethics to the environment fail to provide us with a convincing system for how we ought to behave toward non-human nature.

Given these difficulties, perhaps we should ask why we should even seek an environmental ethic. Most of us have attitudes and values toward nature that are already in place. We – the authors as well as those who share our general outlook – already place a high value on the survival of endangered species and the protection of landscapes, natural features, and natural systems, and we disapprove of activities that needlessly harm non-human entities.

Rather than starting from scratch and asking "How should we act? What should we value?," we seek an environmental ethic that can explain these attitudes and values. We want to give a rational account of our beliefs. If we can rationally and fully justify our values (that is, if we can create an environmental ethic), we can also use reason to persuade others to share these values.

Of course, it may be that such a search is misplaced. It may be that these values simply cannot be justified, any more than, say, someone's acceptance of a particular religion can be justified. If that is indeed the case, we could still try to persuade others to adopt our values, but we would not be able to advance reasons for doing so.

If this is true, and we simply are unable to use the principles of ethics to justify our attitudes toward non-human nature, where does this leave the engineer? How, then, can the engineer incorporate environmental ethics into professional decision-making if there are no rules or principles that can be seen to be universal and rationally defensible?

It is in this spirit of curiosity and necessity that the following chapters search for an environmental ethic.

Suggested Supplemental Readings for Chapter 2

1. *The Existential Pleasures of Engineering,* Samuel Florman
2. *Decision Making in the Corps of Engineers,* P. Aarne Vesilind
3. *Future Generations and Social Contract,* Kristin Schrader-Frechette
4. *Consumption, Conservation, Use, and Preservation,* Alastair S. Gunn
5. *The Tragedy of the Commons,* Garret Hardin

3

The Search for Environmental Ethics in Professional Codes of Ethics

As we argue in the preceding chapter, engineers are typically utilitarians and function as trustees of the public good. Individual members of the public, however, do not make utilitarian decisions when the consequences of the action can adversely affect them personally or other innocent persons. They are thus uncomfortable when engineers make decisions for them on the basis of what the *engineers* think is best, based on utilitarian calculations. Although unsure of how such decisions should be made, much of the public dislikes any method where costs and benefits are unequally distributed or where a price is placed on a human life. This clash of philosophies has made agreement between engineers and the public on the proper role of engineers difficult.

Professional autonomy has always been considered beneficial to the public. If the government starts telling physicians how to treat people, or preachers what to preach, or engineers how to build things, then the public loses. Accordingly, the professions have jealously guarded their autonomy in the name of the public good.

Such autonomy can, of course, be taken away by the state, as witnessed in nations with totalitarian governments. The engineering profession has recognized that the autonomy of engineering depends on the level of public trust, and that it is very much to the advantage of engineering and the public at large to maintain this trust.

To promote such trust and respect, most professional societies have drafted statements that express the values and aspirations of the profession, statements commonly referred to as "codes of ethics." In this chapter we examine these codes, and ask if they provide a basis for making engineering decisions concerning the environment.

3.1 The Engineering Code of Ethics

Engineers have always seen themselves as belonging to a profession, and, not surprisingly, every engineering discipline has its own code of ethics.

One of the earliest codes was adopted in 1914 by the American Society of Civil Engineers. In this discussion we focus on the ASCE Code for several reasons. First, what civil engineers do often has a significant effect on the environment. While the work of other engineers also affects environmental quality, civil engineering is the engineering of constructed facilities, and these facilities invariably alter the environment. Second, the ASCE Code is representative of most of the other engineering codes. With the notable exception of the Code of the Institute for Electronic and Electrical Engineers (IEEE), all engineering codes resemble the Code of Ethics of the Accreditation Board for Engineering and Technology (ABET), the organization that accredits schools of engineering in the United States. Lastly, the leaders of ASCE have recently reconsidered the question of environmental ethics, and their actions are instructive for us.

The ASCE Code has changed frequently over the years. Based in spirit on the original Code of Hammurabi,[1] the 1914 ASCE Code addresses the interactions between engineers and their clients, and among engineers themselves. Only in 1963 did the ASCE modify the Code to include statements about the engineer's responsibility to the general public. But even at this writing the Code does not have an enforceable provision addressing the engineer's responsibility toward the environment, except perhaps in cases where an environmental change would directly and adversely affect the health, safety, and welfare of the public.

But why have a code in the first place? Why do civil engineers, as well as the other professions, want to bother with a code of ethics? Codes of ethics are ubiquitous for three basic reasons: they serve as public relations documents, as contracts among the members of the profession, and as a means of encouraging members of the profession to make decisions in the public interest.

Public Relations

Although not much is known about early professional codes, we do know that medieval guilds codified their rules of conduct to clarify what was considered professional behavior for members of the guild. The underlying purpose of these codes was to enhance the power of the guild and thereby provide wealth and job stability to its members. The guild members, once they adopted the code of conduct, followed it scrupulously because breach

---

1.   Hammurabi, *The Code of Hammurabi, King of Babylonia, About 2250 B.C.*, ed. Robert Francis Hope (Wm. W. Gaunt, Holmes Beach, FL, 1994).

of the rules would mean expulsion from the guild and great personal loss. Guilds in major medieval cities had guild courts that tried their members for misconduct, and often these courts took precedence over lay and ecclesiastical courts. But the guilds grew too powerful and, having defied the public good, were eventually abolished by legislation.[2]

The similarity between professional societies and medieval guilds has been noted by others, including a former president of the now defunct American Institute of Electrical Engineers, who at the turn of the century urged the profession to develop more "guild spirit" or "that force which makes for the increase of prestige, influence, and power of the guild."[3]

Modern professional societies, of course, might someday share the fate of the guilds and have their special privileges revoked by legislation. In response to this fear, and apparently having learned a lesson from the guilds, the modern professions justify their existence and their special privileges by arguing that these are in the best interest of society. They have had to convince the public that it is in everyone's interest to continue the (guild) system.

This public relations effort has been accomplished in part through the development and dissemination of professional codes of ethics. These codes not only state rules of conduct, as did the medieval codes; they also include statements (or platitudes, to the cynical) about "service to the public." For example, the first "Fundamental Canon" in the ASCE Code of Ethics proclaims:

> Engineers shall hold paramount the safety, health and welfare of the public in the performance of their professional duties.

and the first "Fundamental Principle" states:

> Engineers uphold and advance the integrity, honor and dignity of the engineering profession by using their skill for the enhancement of human welfare.

Whenever the government or the public question the motivations of engineers, or the profession itself, the professional society relies on the "service to the public" statement of the Code to ward off undesirable (in the opinion of the engineers) legislation. Although engineers like to be-

---

2.   George Unwin, *The Guilds and Companies of London* (Methuen, London, 1925).
3.   E. T. Layton, Jr., *The Revolt of the Engineers* (Johns Hopkins University Press, Baltimore, 1971).

lieve that the restrictive practices of the profession promote quality control and that the enhancement of professional image is only for the good of the public, some critics believe the sole reason for including these aspirational statements is self-serving public relations. John Kultgen, for instance, argues that the whole discourse of professionalism is no more than rhetoric designed to maintain the elite status and rewards of the members. Critics like Kultgen find even more self-serving rhetoric in the societies' claim that maintaining high status is essential for high-quality service.[4]

Such criticism contains some truth. Public relations and self-interest have been important reasons for the inclusion of aspirational statements in codes of ethics. When professional societies fail in public relations, as the guilds did, the members of that society risk losing status and respect. Aspirational public relations statements assure the public that the profession does indeed have the best interest of society as its paramount purpose. As such, these statements justify to the public the societies's power to police its own members.

It might appear that all this activity is motivated purely by self-interest, and indeed it is quite possible that without such public relations effort, governmental restrictions on practice will be enacted and status and profitability will be eroded. But if this happens, the caliber of persons entering the profession may deteriorate and the quality of engineering suffer, all to the detriment of the public good.

Professional Conduct

A second motivation for adopting a code of ethics, as with the medieval guilds, is the regulation of the professional conduct *for the benefit of its members*. For example, most codes forbid engineers to advertise. By eliminating advertising costs, everyone gains (except the advertisers). Most codes also state that all engineers should be paid on a professional scale commensurate with the salaries of other engineers. Thus, a code of ethics also attempts to establish a union scale for all engineers.

Statements such as the salary requirement have little to do with ethics per

4.   J. Kultgen, "The Ideological Use of Professional Codes," *Business and Professional Ethics Journal*, vol. 1 (1982), pp. 53–69.

se. Nevertheless, they create a professional contract among members and their employers, exactly as the original guild agreements functioned. The code is therefore a mechanism for maintaining and promoting professional profitability by setting down strict regulations. According to John Ladd, "the underlying purpose of the codes is to create some kind of behavior control analogous to control through law."[5]

We therefore conclude that a second motivation for adopting these codes is to ensure conformity to rules of professional conduct. While the aspirational statements reassure the public, the rules control behavior for the benefit of both the profession and its individual members. But again, even these rules should serve the public by promoting a well-regulated profession.

Public Good

Engineering codes of ethics, especially the more recent revisions, also express an altruistic vision of engineering. The key word in the ASCE code (and many other professional engineering society codes) is "paramount." For decisions that involve public health, safety, and welfare, the code of ethics requires the engineer to place *paramount* the benefit of the public.

As discussed in Chapter 2, such provisions for the public good are fraught with difficulty, especially because there is no single definition of "the public." For example, suppose Bruce, the engineer in Case 6, decides he simply cannot ignore the effect the new refinery would have on the environment and thereby the public good, and decides to go to the local press with his report and findings. Under the ASCE Code of Ethics, he would violate both its confidentiality and its loyalty provisions. If pressed, however, he could say that his regard for the public good overrode these professional requirements.

It would be advantageous, of course, if decisions in the public interest always resulted in personal rewards, or at least that they did not require the engineer to be a moral hero. But what happens when carrying out the behavior advocated in the aspirational sections of a code results in financial or professional *loss* to the individual? It is easy to be ethical when such behavior is rewarded, but it is a different matter when the correct action as defined by the code may lead to personal financial ruin or loss of future employment

---

5. J. Ladd, "The Quest for a Code of Professional Ethics: An Intellectual and Moral Confusion," in Chalk et al., *AAAS Professional Ethics Project* (AAAS, Washington, DC, 1980).

opportunity. Unfortunately, the engineering profession is littered with trun-cated careers and financial ruin as the result of engineers making the "right" decision.[6]

Given that engineers inevitably make value and policy decisions, they must do so openly, honestly, and effectively. To some experts, engineers have quite onerous obligations to protect the public interest, for instance, by whistle-blowing. K. D. Alpern believes:

> Engineers have a duty to make personal sacrifices in calling attention to defective design, questionable tests, dangerous products, and so on. . . . The engineer [must] be willing to make greater personal sacri-fices than can normally be demanded of people in general. This quali-fies engineers as moral heroes of a certain sort.[7]

But when exactly is whistle-blowing necessary? Most writers emphasize that it should be used only in situations where there is a serious risk of harm to innocent persons and only as a last resort. Engineers hold this duty because only they have the expertise to recognize the dangers. The products of modern technology are simply too complex and specialized for the aver-age person to understand. Many laypeople, and no doubt many engineers, neither have nor seek the faintest glimmer of understanding of the workings of the personal computers they regularly use, except that if you press cer-tain keys, something appears on the screen. Indeed, the technology is not even visible but is enclosed in a plastic box. Open it up and all you see is a collection of apparently identical components soldered together. Likewise, airplane passengers are quite unable to examine the planes in which they fly, and depend on engineers to detect any cracks. Engineers have that exper-tise – and the high prestige and income that it earns – but the expertise and rewards come at a price. The price is that they must apply that expertise for the public good, even if this may threaten their prestige and income.

Critics of this view argue that it oversimplifies the would-be whistle-blower's ethical situation. Sissela Bok believes that professionals "must weigh the responsibility to serve the public interest against the responsibil-ity owed to colleagues and the employer institution."[8] Others, such as

---

6.   For real-life case studies, see R. J. Baum, and A. W. Flores, *Ethical Problems in Engineering* (Rens-selaer Polytechnic Institute, Troy, NY, Human Dimensions Center, 1978).
7.   K. D. Alpern, "Moral Responsibility for Engineers," *Business and Professional Ethics Journal*, vol. 2 (1983), pp. 39–48.
8.   S. Bok, "Whistleblowing and Professional Responsibilities," *Professional Engineer*, vol. 49 (1979), pp. 26–27.

Richard DeGeorge, go further and argue that whistle-blowing is justified only as a last resort:

> Engineers in large corporations have a important role to play. That role, however, is not usually to set policy or to decide on the acceptability of risk. Their knowledge and expertise are important both to the companies for which they work and to the public. But they are not morally responsible for policies and decisions beyond their competence and control.[9]

Given the uncertainty among experts on whistle-blowing, the engineering codes of ethics play a helpful role by supporting whistle-blowing that is in the public good. Thus, the third function of engineering codes of ethics is to encourage engineers to act in the public good. Indeed, most engineers believe that the profession *is* a service profession and that the professional societies were created to promote service to the public. As the societies' position on whistle-blowing indicates, these professional codes of ethics reflect more than shrewd public relations. They represent a legitimate desire to encourage engineers to act ethically.

In summary, the underlying motivations for adopting a code of ethics in professional engineering are:

1.   to define ideal behavior for the purpose of enhancing the public image,
2.   to establish rules of conduct for policing its own members, and
3.   to encourage value-laden decisions for the public good.

3.2   Development of the ASCE Code of Ethics

As an illustration of the function of engineering codes of ethics, consider the development of one of the earliest and most important (relative to environmental quality) engineering codes of ethics – that of the American Society of Civil Engineers.

The ASCE has always been the most restrictive of the original engineering societies. Its membership is limited to graduates of accredited engineering schools and those who have the recommendation of two ASCE mem-

---

9.   R. T. DeGeorge, "Ethical Responsibilities of Engineers in Large Organizations," *Business and Professional Ethics Journal,* vol. 2 (1981), pp. 1–6.

bers. One does not *join* the ASCE – one is "elected." Upon election, the engineer must sign a pledge to abide by the ASCE Code of Ethics.

The ASCE has also developed a complex and effective mechanism for enforcing the Code. If a member of the ASCE is convicted at a hearing (resembling that of a medieval guild court) of a serious breach of the ASCE Code of Ethics, the member can be expelled from the Society. Such an action would not necessarily affect the engineer's professional engineering (PE) license, which is regulated through the individual states, and the engineer can continue to practice engineering. Conversely, the loss of the PE license would not directly affect his/her standing in the Society.

Not all civil engineers take the time or trouble to belong to the ASCE. There are perhaps 2 million civil engineers in the United States, defined either by training or occupation, and only about 100,000 of these are members of the ASCE. For the remainder, the ASCE Code of Ethics has no legal or moral significance. In the United States, at least, it is impossible to prevent a civil engineer from practicing engineering, regardless of the reason. Neither membership in ASCE nor a professional engineering license is necessary, and thus the loss of ASCE membership or the PE license will not prevent engineers from working in their areas of technical expertise. The only constraint is that if engineers are not registered (do not have a professional engineering license), they cannot be in responsible charge of projects where public money is involved. As a result, the vast majority of engineering graduates never bother getting professional engineering licenses or joining the ASCE.

But civil engineers in responsible charge of projects are required by law to be registered PEs.[10] The logic for this is that civil engineering involves the construction of public facilities such as roads, bridges, buildings, and water supplies. Since civil engineers in responsible charge are always registered, and almost always members of ASCE, their Code of Ethics has a central role in the conduct of the profession and its purpose and origin are an important part of civil engineering.

While personal integrity has always been a part of engineering, the adoption of a formal code is a relatively modern development. At the time of

---

10. "Responsible charge" has a legal meaning. For every project, one engineer is responsible for its completion, If it fails, it is the responsibility of that engineer. He or she cannot blame others, since it is the engineer's responsibility to catch all of the mistakes. In some European countries, an engineer in responsible charge of designing a bridge was required to stand under the bridge while it was being load-tested.

ASCE's founding in 1852, some members wanted a code of ethics, but the Society instead adopted the position that "professional ethics [are] considered to be a strictly personal responsibility – a matter of honor."[11] The Society reinforced this stance in 1877 when it considered a proposal regarding what members should do when non-engineers overruled their engineering judgment. As in 1852, the ASCE adopted a position of aloofness:

> Resolved: That it is inexpedient for this society to instruct its members as to their duties in private professional matters.[12]

The first formal proposal for a code of ethics came in 1893 when a group of members from Cincinnati called for the appointment of a committee to draft such a code. The Society took no action on this matter and dropped the idea of a code.

The idea of a code of ethics resurfaced in 1902, but once again the ASCE decided against it. In the words of the Society's president,

> . . . it is, I think, safe to say that for the kindling of professional enthusiasm, and the establishment of high professional standards, the Society and its members will continue to rely, as they have done in the past, upon these vital and moral forces, and not upon the enactment of codes or upon any form of legislation.[13]

The Society felt new pressure to adopt a code of ethics again in 1913 when the upstart American Society of Mechanical Engineers adopted its own code. The ASCE board of directors appointed a committee to review the matter, and finally, in 1914, the membership adopted its first Code of Ethics.

The original Code was a simple list of rules for professional conduct between engineer and engineer, and engineer and client – essentially nothing more than a set of guild regulations. Nevertheless, the ASCE Code differed from other engineering society codes in that it became perhaps the most effective enforcement instrument of any professional society in the United States.

The adoption of the Code of Ethics did not, of course, change the basic tenor of ASCE overnight. When in 1920 a group of young engineers tried to include aspirational articles in the Code, the rest of ASCE soundly thrashed them. The power and prestige in ASCE remained with senior engineers in

---

11. W. H. Wisely, "Professional Turning Points in ASCE History," *Civil-Engineering* (October 1977).
12. Ibid.
13. Ibid.

New York, who knew very well how to conduct their own business. Finally, after several scandals that threatened to result in legislative action, the ASCE revised its Code to include the public-good-is-paramount clause.

The present ASCE Code of Ethics, reproduced in the supplemental readings to this chapter, has three parts: *The Fundamental Principles, The Fundamental Canons,* and *The Guidelines to Practice.*

The Fundamental Canons, of which there are seven, are the revised form of the original 1914 Code of Ethics. The Guidelines to Practice, first published in 1961, clarify and add detail to the Fundamental Canons and appear as an attachment to the Code.

The Fundamental Principles appeared in 1975. Their origins are unclear but no doubt are related to the notorious Spiro Agnew affair, which gave civil engineering a deserved black eye.[14] The Fundamental Principles were copied verbatim from the Code of Ethics of the Accreditation Board for Engineering and Technology. Whereas the Fundamental Canons contain specific ethical statements, the Fundamental Principles are broader in nature and relate to the engineer's role in society.

Some interesting ASCE debates about the Code are worth mentioning here. In 1963, for example, the Society debated the "when in Rome" clause, which was eventually adopted as a footnote to the Code. It read:

> On foreign engineering work, for which only United States engineering firms are to be considered, a member shall order his practice in accordance with the ASCE Code of Ethics. On engineering works in a foreign country he may adapt his conduct according to the professional standards and customs of that country, but shall adhere as closely as practicable to the principles of this Code.

This provision allowed United States engineers to use bribery in countries where such practices were tolerated. Eventually the footnote was repealed, amid a great debate over the role of the American engineer in international work.

---

14. Spiro Agnew, a member of the Baltimore County Board of Commissioners, used his position to funnel roadwork to engineers who, in turn, provided him suitable kickbacks. When Agnew become governor of Maryland, he continued this practice on a larger scale. When he became Richard Nixon's vice-president, Agnew continued to collect the kickbacks, some of which were delivered by the engineers in plain brown envelopes to the office of the vice-president of the United States. Needless to say, the affair was an acute embarrassment to civil engineering. When the whole sordid affair came to light, Anew resigned from the vice-presidency. If he had not done so, he could have become the president of the United States when Richard Nixon resigned.

A second brouhaha occurred in 1972 when the Department of Justice wanted to sue ASCE for restraint of trade over a section of the Code that restricted competitive bidding. The ASCE defense was that when the low bidder designs public works, the quality suffers and the public good is impaired. The Department of Justice held that if there is no market constraint on the cost of engineering design, the profession in effect becomes a monopoly. After much discussion and breast-beating, the ASCE agreed to remove the constraint from its Code, marking the only time that a public agency had directly influenced the ASCE Code of Ethics.

ASCE and Environmental Ethics

For all its merits, the present ASCE Code of Ethics is a collection of "do's" and "do not's" concerning *human* interaction – engineer/engineer, engineer/client or engineer/public. The Code only grudgingly and ambiguously recognizes the involvement of engineers in environmental matters, long a controversial aspect of engineering.

Responding to the general growth of environmental awareness and conscious of the popular image of civil engineering as the perpetrator of environmental destruction, the ASCE Code was revised in 1977 to include the following statement:

> 1.f   Engineers should be committed to improving the environment to enhance the quality of life.

Note first that "Engineers *should*. . . . " is very different from "Engineers *shall*. . . . " The use of "should" in effect precludes the enforcement of this section of the Code. All enforceable sections begin with the statement "Engineers shall. . . . " Further, note that environmental effects relate solely to quality of life. Although the Code is vague on the matter, the phrase "quality of life" presumably applied only to *human* life. The Code in no way suggests that nature has intrinsic value beyond its utility, or instrumental value, to humans.

The next attempt to bring environmental considerations into civil engineering practice was the adoption by the board of directors of Policy Statement No. 120, which outlines ASCE policy concerning environmental protection.[15] The key phrase in the policy statement is

---

15.   "Assuring a Desirable Quality of Life," ASCE Policy statement No. 120 (New York, April 1980).

. . . ASCE recommends that the individual civil engineer dedicate himself or herself to the following objectives.

The objectives are:

1. Civil engineers must recognize the effect their efforts will have on the environment, by increasing knowledge and competence in incorporating ecological considerations in design.

2. The civil engineer must inform a client of the environmental consequences compared with the benefits of the services requested and the design selected, recommending only responsible courses of action.

3. The civil engineer must fully utilize mechanisms within the Society which lend support to individual efforts to implement environmental considerations.

4. The civil engineer must recognize the urgent need to take the lead in development, modification and support of efficient government programs, to insure adequate environmental protection but avoid the inhibition of the economy which can result from overregulation.

As a comprehensive environmental policy for civil engineers, the ASCE policy statement leaves much to be desired. Note that even though the four articles use the word "must," the entire policy is only a "recommendation," thus effectively negating any regulatory or even moral function. The third objective is the most curious. Nobody seems to know what is meant by "fully utilize mechanisms within the Society which lend support," especially because the objectives are not enforceable by the ASCE. Finally, the fourth objective clearly puts short-term economic benefits above long-term environmental costs.

Recognizing this deficiency in the Code, the Environmental Impact Analysis Research Council (EIARC) of the Technical Council on Research (a committee of the ASCE) proposed in 1983 an eighth fundamental canon.[16] The proposed canon reads:

8. Engineers shall perform service in such a manner as to husband the world's resources and the natural and cultured environment for the benefit of present and future generations.

---

16. The Environmental Impact Analysis Research Council of Technical Council on Research, "A Proposed Eighth Fundamental Canon for the ASCE Code of Ethics," *Journal of Professional Issues in Engineering*, vol. 110, no. 3 (July 1984).

Listed under the canon are nine guidelines that elaborate the canon. For example, guideline 8.g reads:

> Engineers, while giving proper attention to the economic well-being of mankind and the need to provide for responsible human activity, *shall* be concerned with the preservation of high quality, unique and rare natural systems and natural areas and *shall* oppose or correct proposed actions which they consider, or which are considered by a reasonable consensus of recognized knowledgeable opinion, to be detrimental to those systems or areas. (emphasis added)

The proposal struck many people as relatively modest and uncontroversial. It is explicitly anthropocentric: the environment is to be protected "for the benefit of present and future generations" – of humans, obviously. Nonetheless, in their January 1984 meeting the Professional Activities Committee voted *unanimously* to *not* recommend approval, and the canon died there.[17] The reasons for this rejection noted in the minutes were that environmental concerns are adequately covered by article 1.f (just quoted) and Policy Statement No. 120. Members of EIARC, which first drafted this proposal, were told that it was *legal considerations* that prompted the disapproval of the canon. A former chairman of EIARC later admitted that this reason seemed implausible.[18]

Some enlightened members of ASCE decided to try an end run and created a different committee, the Committee on Engineering Responsibility (COER), and assigned it the task of getting the eighth canon approved. The committee met several times and was about to submit a revised canon for approval when suddenly the committee was disbanded. For ASCE this appeared to be the simplest way of rejecting the troublesome eighth canon without going on record as opposing it.

Why are the leaders of ASCE so scared of this revision to the Code of Ethics?

The answer becomes clearer if we remember the motivations for the entire Code of Ethics. Recall that professional codes are written to:

1. Enhance public image
2. Provide rules for enforcing conduct
3. Promote public welfare

---

17. Minutes of the Professional Activities Committee of ASCE, January 1984.
18. Personal communication.

The policy makers in the ASCE assumed that Policy Statement No. 120 would protect ASCE's public image adequately. But what about the two remaining objectives? Would the proposed canon have provided rules that the ASCE would have wished to enforce? And would it have prompted decisions truly in the public interest?

Consider what might occur to an engineer accused of unwarranted environmental destruction if the eighth canon became a part of the Code. The Code would allow the ASCE to prosecute the engineer and to encourage engineers on the job to become active whistle-blowers on behalf of the environment. Useless canals through pristine forest, airports in wetlands, and roads through wildlife sanctuaries would all become more difficult to construct. Short-term professional profitability would no doubt suffer, and the "inhibition of the economy" on behalf of environmental protection might actually occur.

Of course, one can argue that all construction is in the public interest. If this is true, then the ASCE Code certainly promotes the public good without the controversial eighth canon. But if some construction, despite its short-term economic benefits, is actually *detrimental* to the public good, then the defeat of the proposed canon was a defeat for the public. Without such a canon, engineers hold no professional obligation to concern themselves with troublesome environmental ethics.

The Sustainable Development Alternative

Seeking alternatives to the troublesome eighth canon, ASCE discovered "sustainable development." This term seems, on the surface, to be an oxymoron. How can development go on forever? We recognize that the earth cannot sustain limitless growth, so how can we sustain ceaseless development?[19]

The term sustainable development was first popularized by the World Commission on Environment and Development (also known as the Brundtland Commission), sponsored by the United Nations. Within this report sustainable development is defined as "development that meets the needs of the present without compromising the ability of future generations to meet their own needs."[20] There are very many definitions of the term – the

---

19. For this reason, the New Zealand Resources Management Act refers to sustainable *management* instead of *development*, as discussed in Chapter 7.2.
20. World Commission on Environment and Development, *Our Common Future* (Oxford University Press, Oxford, 1987).

Brundtland Report itself includes a further ten – while a report for the United Kingdom Department of the Environment contains thirteen pages of definitions.[21]

Although the original purpose of this term was to recognize the rights of the developing nations in using their resources, sustainable development has gained a wider meaning and now includes educational needs and cultural activities, as well as health, justice, peace, and security.[22] All these are possible if the global ecosystem continues to support the human species. We owe it to future generations, therefore, not to destroy the earth they will occupy. According to the World Bank:

> The sustainable approach to development . . . contains a core ethic of intergenerational equity, along with an understanding that future generations are entitled to at least as good a quality of life as the present ones.[23]

Most engineers would subscribe to this ideal. But engineers deal in operations – they are doing things – and therefore need an *operational* definition of sustainable development. Some have suggested a modified environmental impact analysis, not only taking into account the effect on the present environment, but also considering effects on future generations and global ecosystems. But these techniques have the same problems as the original environmental impact studies – they depend on crossover valuation of incompatible goods. What is more valuable, our present needs or the needs of future generations? How should the needs of wildlife for forests be balanced with the need for lumber? What pain and suffering by a laboratory test animal do we accept in order to reduce health problems in humans? And who is to decide these questions?

Clearly the ideal concept is a long way from an operational definition. Still, with all its problems, sustainable development is a worthwhile idea. Recognizing this as a positive step in defining the responsibilities of engineers toward the environment, the First Fundamental Canon in the 1997 revisions of the ASCE Code of Ethics was changed to read:

---

21. D. Pearce, A. Markanya, and E. B. Barber, *Blueprint for a Green Economy*, Report for the UK Department of the Environment (Earthscan Publication, London, 1989).
22. Joseph R. Herkert, Alex Farrell, and James Winebrake, *Technology Choice for Sustainable Development*, IEEE Technology and Society Magazine, vol. 15, no. 2 (Summer 1996).
23. J. Pezzy, *Sustainable Development Concepts: An Economic Analysis* (World Bank Development Paper No. 2, Washington, DC, 1992).

1. Engineers shall hold paramount the safety, health and welfare of the public and shall strive to comply with the principles of sustainable development in the performance of their professional duties.

At the surface, ASCE has taken a giant step forward in incorporating environmental values into its Code of Ethics. But let's look at this more closely. Consider the wording. The engineer shall (that's a good start) strive (meaning that the engineer has to try, not actually do) to comply with the principles of sustainable development. But nowhere in the Code are the principles of sustainable development spelled out.

Principles of sustainable development are not like the laws of thermodynamics, or regulations on stream quality, or traffic laws. Engineers are free to determine what in their opinion are the principles of sustainable development, and then all the Code asks of them is to strive to act so as to be in line with what they themselves determine to be these principles.[24]

Worse, under the *Guidelines to Practice Under the Fundamental Canons of Ethics* paragraph 1f. reads:

1f. Engineers should be committed to improving the environment by adherence to the principles of sustainable development so as to enhance the quality of life of the general public.

Ignoring the curious reference to the "general" public (who else is there?), the key word is, of course, should. Even though the Fundamental Canon says shall, the guideline lets the engineer off the hook by suggesting that should is good enough.

The cynic would say that once again the American Society of Civil Engineers has changed its Code of Ethics to enhance its public image and not to effect a meaningful change in the actions of civil engineers. But this is unkind. We believe that engineers are seriously trying to cope with environmental problems and balancing the rights and benefits of humans and the non-human environment. There are many sensitive and caring civil engineers who want to do the right thing in the performance of their duties, and some of them no doubt were instrumental in effecting the change in the Code of Ethics. But as important as this first step is, the engineer who seeks guidance in the ASCE Code of Ethics for making decisions that affect the environment will be disappointed to find little of useful value.

---

24. By contrast, the Institute of Professional Engineers in New Zealand and the Institution of Engineers in Australia have developed quite detailed policies on sustainable development (Australia) and sustainable management (New Zealand).

3.3 Conclusions

Sadly, the ASCE Code of Ethics, which is representative of other codes adopted by mainstream American engineering societies, does not help us in making decisions that involve environmental values. Further, the leadership of the ASCE appears to be incapable and/or unwilling to fix this defect. If we are to find guidance on how to approach and solve engineering problems related to the environment, we must search elsewhere.

A few years ago a group of international engineers came to essentially the same conclusion, and decided to write a comprehensive code of environmental ethics. This document, reproduced here, makes interesting reading; and it is even more interesting to speculate on the possibility of its acceptance by organizations such as the American Society of Civil Engineers.

Code of Environmental Ethics for Engineers [25]

The WFEO COMMITTEE ON ENGINEERING AND ENVIRONMENT, with a strong and clear belief that man's enjoyment and permanence on this planet will depend on the care and protection he provides to the environment, states the following principles.

TO ALL ENGINEERS

When you develop any professional activity:

1. Try with the best of your ability, courage, enthusiasm and dedication to obtain a superior technical achievement, which will contribute to and promote a healthy and agreeable surrounding for all men, in open spaces as well as indoors.
2. Strive to accomplish the beneficial objectives of your work with the lowest possible consumption of raw materials and energy and the lowest production of wastes and any kind of pollution.
3. Discuss in particular the consequences of your proposals and actions, direct or indirect, immediate or long term, upon the health of people, social equity and the local system of values.
4. Study thoroughly the environment that will be affected, assess all the impacts that might arise in the state, dynamics and aesthetics

---

25. Approved by the Committee on Engineering and Environment of the World Federation of Engineering Organizations, in the 6th Annual Plenary Session, New Delhi, 5 November 1985. Published in Buenos Aires, WFEO Headquarters, 1986.

of the ecosystems involved, urbanized or natural, as well as in the pertinent socio-economic systems, and select the best alternative for an environmentally sound and sustainable development.

5. Promote a clear understanding of the actions required to restore and, if possible, to improve the environment that may be disturbed, and include them in your proposals.

6. Reject any kind of commitment that involves unfair damages for human surroundings and nature, and negotiate the best possible social and political solution.

7. Be aware that the principles of ecosystemic interdependence, diversity maintenance, resource recovery and interrelational harmony form the bases of our continued existence and that each of those bases poses a threshold of sustainability that should not be exceeded.

> Always remember that war, greed, misery and ignorance, plus natural disasters and human induced pollution and destruction of resources, are the main causes of the progressive impairment of the environment and that you, as an active member of the engineering profession, deeply involved in the promotion of development, must use your talent, knowledge and imagination to assist society in removing those evils and improving the quality of life for all people.

Suggested Supplemental Readings for Chapter 3

1. *The Philosophical Basis of Engineering Codes of Ethics,* Albert Flores
2. *The Quest for a Code of Professional Ethics: An Intellectual and Moral Confusion,* John Ladd
3. *What Are Codes of Ethics For?,* Judith Lichtenberg
4. *The Code of Ethics* of the American Society of Civil Engineers

4

The Search for Environmental Ethics by Applying Classical Ethical Theories

A fundamental principle of ethics in modern liberal-democratic societies is that it is to everyone's overall advantage to act morally. Civilized society is built on a tenuous foundation. The benefit of all requires the acceptance and cooperation of all. Once this spirit of cooperation breaks down, it is difficult if not impossible to reestablish. Consider the hatred among different groups in the Middle East, the Balkans, Ireland, or even the inner-city United States.

All long-standing societies have a shared culture that is central to the continuation of the society, including certain shared values that members of the society accept and follow. A shared sense of how we should act, the acceptance of civility, and caring neighbors are requisites for a desirable social order.

We may speak of members of a society as *knowing* that certain acts are right or wrong, both because they know what the rules are and because they accept that the rules apply to them along with everyone else. Inevitably, especially in pluralistic societies where there is disparity in well-being, not everyone subscribes to the theory or practice of what we call the core values of the culture. Nonetheless, most people agree, for example, that stealing from each other is wrong, so wrong that we create laws to punish thieves. We also accept that it is wrong to break into line at the checkout counter. Although we have no laws for checkout counter behavior, and the guilty certainly do not risk jail by breaking into line, we still agree to respect others in the line, and teach our children to behave accordingly.

Similarly in engineering, we recognize that clients have a right to expect confidentiality and loyalty from engineers. Without this trust, inefficient, inappropriate, and even unsafe structures/products/processes may result. It is to everyone's advantage to honor client confidentiality, and this is recognized in the engineers' code of ethics.

Many of the questions concerning values are clear-cut. Do not steal. Do not break into line. We do not need any further explanations or contemplations about such issues. If all decisions were that simple, there would be no need for ethics. Unfortunately, few problems in life are simple and we need to find a way to help us think about conflicting choices.

Although we know that stealing is wrong, perhaps stealing to save a friend's life is acceptable. But is it acceptable if what is stolen will result in the death of another human being?

Although we know that breaking into a line at a checkout counter is wrong, perhaps this is acceptable if the lines are long and the purchase is a medicine that can stop the pain of a hurting child.

Although we know it is good to retain client confidentiality, should this be the overriding value when the client is about to pollute a vast groundwater resource and refuses to reveal the problem to the regulatory agencies? Should Peter in Case 1 report the old oil spill to the authorities? Would it matter if the contaminated groundwater is not used as a human freshwater source for drinking purposes?

When confronted by such sticky questions, we must stretch and rethink our basic values. We of course have to make decisions, but how can we know if these decisions are the correct ones? How can we know that we are acting ethically? Perhaps the wisdom of our classical ethical traditions can be of use. Perhaps these can teach us how to think through these problems and to provide systematic techniques for the resolution of ethical dilemmas.

In this chapter, we present an essential introduction to some of the major themes in Western ethical theory. We then discuss how ethical theories have approached environmental questions, and ask how these ideas can help engineers make ethical decisions concerning the environment.

We do not claim that these, or any other theories, apply to all societies and for all times. Whether there are such universal values is a highly controversial question. For instance, in most cultures, including premodern European cultures, individual rights are not ethically central, and other values such as community, harmony, fairness, or spiritual development may be central.

However, we do not wish to pursue such issues here. Instead, we explore the central values of Western ethics in order to evaluate the possibility of an environmental ethic based on these ethical theories.

4.1 Ethical Theories

An ethical theory is an attempt to answer certain questions about ethics. Ethical theories do not directly provide answers to specific questions such as whether to allow prayers in public schools or how much, if anything, to donate to charity. Rather, they decide what moral principles should be adopted. In the Western tradition, ethical theories are of two main types – consequentialist (or teleological) and deontological.

Consequentialist Theories

Consequentialist theories evaluate acts, policies, practices and institutions according to their consequences. Stated simply, in such theories a right action is one that overall has good consequences; a wrong action is one that overall has bad consequences.

The most influential consequentialist theory is *utilitarianism,* developed in great part by English legal theorist and philosopher Jeremy Bentham (1748-1832). His view, often referred to as *classical utilitarianism,* is that only happiness, which he defines as pleasure and the absence of pain, is good in itself. Everything else is good only as a means to happiness. Thus exercise, disagreeable though it may be, is good because it is a means to health, and health is good because healthy people are happier than unhealthy people.

For utilitarians, an act is right in proportion as it tends to increase the sum of human happiness or decrease unhappiness and is wrong if it tends to the reverse. Some philosophers have tried to extend utilitarianism to animals, as we see in Chapter 5.

Modern utilitarians usually refer to preferences and/or interests rather than happiness, but this is not really a significant difference, because the classical utilitarians believed that happiness is in everyone's interests, that everyone desires to be happy, and that happiness either consists in or follows from maximizing preference satisfaction. Most current decision strategies (game theory, cost-benefit analysis, risk-benefit analysis, and other derivatives of operations research) trace their origins to utilitarianism.

Other consequentialist ethical theories locate the "good" elsewhere: for example, *ethical egoism* (the good is that which benefits me), *nationalism* (the good is that which advances the state), or *altruism* (the good is that which benefits others).

Deontological Theories

Deontological theories are nonconsequentialist; deontologists deny that the rightness or wrongness of acts or rules is reducible to the value of their consequences. Deontological theories hold that certain actions are right or wrong regardless of their consequences. The Ten Commandments, for example, represents a well-known deontological system.

The most influential deontological thinker was German philosopher Immanuel Kant (1724–1804). Kant emphasized the absolute value of persons, who as free, rational beings must always be treated as ends in themselves. To act morally is to follow universal moral principles that require respect for persons. Kant argued that a test of the rationality of a moral principle is whether or not it can be *universalized*. In his view, rational persons would not be prepared to adopt a rule for themselves unless they were prepared to accept it as applicable to and by *all* persons.

Kant's theory of the universalizability of ethical judgments is but one version of what he calls the *categorical imperative*. That is, some acts such as truth telling are categorically imperative for all persons, regardless of the situation or the consequences. Because these acts are always required, they can be universalized, or expected of all persons at all times. If you choose to act in a certain way, you must then also agree that you must allow others to act in a similar manner.

Kant argues that certain acts such as lying are *always* wrong, even if you have to lie in order to save the life of another person. He believes that a right action is one done out of "good will" or a respect for the moral law. The good person does right *because* it is right, and not for any other reason.

A contemporary deontological philosopher of morality is Dartmouth College's Bernard Gert, who assumes that rational, impartial persons would agree on a common code that is to everyone's advantage, and that it is rational to be moral. In fact, Gert believes it would be irrational to argue that such a code not be accepted and that it would be irrational to break these rules without adequate justification.[1] Gert's proposed new "ten commandments" are

1. Don't kill.
2. Don't cause pain.
3. Don't disable.
4. Don't deprive freedom.
5. Don't deprive pleasure.
6. Don't deceive.
7. Don't break promises.
8. Don't cheat.

1. Bernard Gert, *Morality* (Oxford University Press, New York, 1988).

9. Don't disobey laws.
10. Don't fail to do your duty.

As Gert of course recognizes, to make such a list invites the objection that one may be in a situation where several of the commandments conflict. Suppose you are late for an appointment, should you speed in your car (disobey the law) or be late (break a promise)? Gert suggests that we ought to follow these rules unless rational, impartial persons would agree that one of them should be broken. In the above example, most rational, impartial persons would agree that speeding merely to meet a promised deadline is unacceptable.

Acts and Rules

Both the consequentialist and deontological ethical theories can be divided further into *act* and *rule* theories. An act theory requires the agent to evaluate each action individually. Thus an *act utilitarian* considers the consequences for human happiness with each choice; and an *act deontologist* tries to discover the intrinsic rightness or wrongness of each possible action. Some act deontologist theories stress the uniqueness of each occasion of choice, emphasizing the importance of the individual's life situation and relationships, the historical timing, and so on.

Rule theories, by contrast, evaluate types or classes of acts rather than acts themselves. *Rule utilitarians* create a system of rules designed to promote human happiness or increase preference satisfaction. With this system of rules, rule utilitarians hope to simplify decision making, because obviously we cannot calculate all the possible effects of each proposed course of action. Rule utilitarians also argue that the existence of rules benefits society by providing order and stability. *Rule deontologists* likewise regard *types* of actions as right or wrong. Instead of considering anew the morality of each act, rule deontologists (and rule utilitarians) create a set of principles to guide their actions. Extreme rule theorists, such as Kant, believe that these principles are absolute and exceptionless. For instance, we must always tell the truth; we do not need to decide whether to lie in a particular situation. Of course, problems would be created when two rules pull in different directions.

Rights Theories

The dominant deontological theory in the United States and most other Western countries for more than two centuries has been the theory of *rights*.

So entrenched is this view of ethics, it is frequently taken for granted that this is what ethics is basically about: for instance, that a topic such as abortion must be discussed only in terms of the rights of the fetus ("right to life") and of the pregnant woman ("right to choose"). A utilitarian approach to abortion, by contrast, would consider the interests of all people involved in the decision.

The English philosopher John Locke (1632–1704) developed a comprehensive theory based on the then novel notion of natural rights. To Locke, all humans are born with inalienable rights that cannot be taken away from us. The most important of these rights, according to Locke, are the rights to life, liberty, and property. Unlike conventional or legal rights, which are bestowed on people by other people, natural rights exist regardless of social acceptance. Locke believes that natural rights are bestowed by God – a view shared by the framers of the United States Constitution – and modern secular claims of human rights are firmly in the Lockean tradition.

Ethics, especially in a legal and political context, particularly emphasizes the mutual recognition and respect of natural rights. However, the security of these rights depends on organized society and government. Ethics therefore requires a "social contract" to protect natural rights. According to Locke, we exchange our "natural" state of anarchy for the guaranteed liberty and security of society. In Locke's view, a rational being will want to live in an ethical society.

Locke's point is valid. Imagine how cruel and miserable our lives would be if we could not count on the honesty and goodwill of our neighbors. Because we all benefit from this social contract, it is our ethical responsibility to uphold it.

Even in Locke's time, the notion of a literal, historical social contract that marked our emergence from a supposed precivilized state was not widely accepted. But the moral ideal behind the social contract – the notion that social organization should rely on mutual agreement and not force – is a powerful idea in ethics. Modern philosophers have used it in the formulation of new theories of ethics. For instance, John Rawls has developed a theory of the just society that emphasizes respect, impartiality, rationality and equality.[2] Rawls asks us to imagine a presocial "original position" in which everyone tries to agree on rights and duties before knowing what their position in society will be. Imagine that under this "veil of ignorance," people

---

2.   J. A. Rawls, *A Theory of Justice* (The Belknap Press of Harvard University Press, Cambridge, MA, 1971).

don't even know their race, nationality, gender, or the period of history when they will live. From this fanciful scenario, Rawls argues that rational people, in forming a social contract, would agree on basic principles of justice: basic liberty for all, and distribution of inequalities to the greatest benefit of the least advantaged.

Other Ethical Theories

Many philosophers have criticized both the consequentialist and the deontological theories as being too restrictive, and have developed alternative approaches to ethics. One such alternative ethical theory is called *situation ethics,* which offers a Christian alternative, basing the ethical responsibility on the agent; right actions are those that one does out of love for others.

Existentialists criticize Kant's universalism and focus on the uniqueness of each choice, and on the responsibility of individuals to make their own moral choices. The value of choice, for the existentialist, lies in its courageous assertion of humanity and autonomy in a meaningless universe. Existentialists are fascinated by situations in which conventional morality fails as a guide to action because it offers several conflicting solutions or offers only equally desirable (or undesirable) alternatives.

In recent years, Western ethical theorists have become increasingly interested in *virtue ethics:* the view that ethics is not a matter of bringing about good consequences, or of carrying out duties, but rather is one of developing a particular character, of becoming a particular kind of person. Most such accounts are objectivist – that is, they claim that *everyone* ought to aim to develop the same set of virtues, with appropriate allowance being made for different circumstances.[3] Aristotle presented a lengthy list of virtues, each of which he characterized as a mean position between two undesirable extremes. Thus, the courageous person is neither cowardly nor foolhardy; the generous person steers a course between profligacy and stinginess; the temperate person is neither an ascetic nor a glutton. There will also be virtues specific to particular kinds of people, such as members of a profession.

Aristotle believed there is a strong connection between virtue and happiness. The virtuous person, in addition to benefiting others, will also neces-

---

3. See in particular A. McIntyre, *After Virtue* (University of Notre Dame Press, Notre Dame, IN, 1981).

sarily be happier in proportion to his or her own good character. Thus, if everyone were virtuous, everyone would be happy.

This century has seen the development of *feminist ethics,* and its development has influenced many other areas in applied ethics – particularly environmental ethics. As there are many varieties of feminism, there is not a single feminist ethic. Some feminist philosophers (and some more "traditional" philosophers such as eighteen-century French thinker Jean-Jacques Rousseau, 1712–1778) believe that men and women are inherently very different, that they think and experience the world in different ways and have different emotional and spiritual lives. As we see in Chapter 6, some feminists have linked male domination of women with environmental destruction. Writers such as Mary Daly believe that women are naturally superior to men, ethically, because they are inherently cooperative and caring rather than aggressive and competitive.[4] Thus, a nurturing role is "natural" for women.

If men are indeed incapable of developing or following a feminist ethic, then there could be no such thing as *engineering* ethics because there are both male and female engineers. However, by no means do all feminists believe that the feminist ethic has a biological basis. What feminist ethics have in common is the belief that conventional approaches are too abstract and rationalistic. Feminist ethicists typically emphasize the significance and uniqueness of individual relationships and the importance of concerns and caring as a response to the needs of others, and an ethic of care is often presented as a basis for professional practice, especially in occupations such as nursing, counseling, and education.[5] Although engineering deals with the material conditions of life, its ethics need to reflect its motto – a "people-serving profession." Feminist ethics can provide a useful reminder that "the client" and "the public" are not just abstractions but real, unique people, for whose benefit engineering exists.

4.2   An Evaluation of Ethical Theories

Philosophers often use examples to illustrate distinctions among different theories. Here we use criminal behavior such as kickbacks paid by engineers

---

4.   Mary Daly, *Gyn/Ecology: The Metaethics of Radical Feminism* (Beacon Press, Boston, 1976).
5.   See, for example: Nell Noddings, *Caring: A Feminine Approach to Ethics and Education* (University of California Press, Berkeley, 1978); Megan-Jane Johnstone, *Bioethics: A Nursing Perspective* (W. B. Saunders/Bailliere Tindall, Sydney, 1989).

to secure work. In a kickback scheme, an engineer agrees to pay a public official some fraction of the profit from a construction project in return for being selected for the project.

Most people in Western societies believe that any criminal law such as the prohibition against kickbacks should punish lawbreakers, but they differ about the justification for punishment. Deontologists believe in *retribution*, that criminals deserve to suffer solely because they have broken the law. By breaking the law, the argument runs, criminals act unjustly, benefit themselves at the expense of others, and fail to respect their fellow citizens. Therefore, society ought to punish criminals. In contrast, consequentialists see punishment in terms of the good it will achieve: that is, *deterrence*. Punishment makes sense only if it results in the protection of society or the offender's reform and rehabilitation. Consequentialists support punishment only if it has good effects. Retributivists may welcome good results, of course, but only as side benefits.

These issues often arise in discussions about environmental quality. In instances of severe environmental damage, such as the Love Canal or the Kepone contamination cases (see the Supplemental Readings), retributivists emphasize the injustice of making the public suffer because of the improper disposal of hazardous wastes. They demand harsh penalties so that chemical companies pay for their alleged misdeeds. Utilitarians, concerned with public welfare, concentrate rather on compensating victims and deterring future would-be polluters.

Consequentialists accuse deontologists of divorcing ethics from human welfare. They argue that the point of ethics is to benefit humans. To consequentialists, a system of ethics that requires us to perform acts without reference to or regardless of their consequences is therefore irrational. The Scottish philosopher David Hume (1711–1776), for example, attacked what he calls the "monkish virtues" of chastity, self-denial, humility, and so on as causing misery without promoting anyone's well-being.

Utilitarians claim that their theory has an empirical, practical foundation because it identifies ultimate value with people's actual values. They argue that claims of the absolute rightness or wrongness of, say, telling the truth or murder have no basis beyond the alleged intuitions of deontologists. If challenged, the latter can do little except assert their faith in absolute values. As we shall see, however, Kantians claim that deontological ethics has a rational basis.

Nineteenth-century utilitarians noted that their deontological opponents often made appeals to some supposed ethical authority, such as the Church or tradition. Such appeals were, and are, unreasonable. To say, "You ought to do it because the Church (or tradition) says so," is to invite the question "Why should I obey the Church (or tradition)?" That some people believe they should live by the teachings of a religion does not in the least bind non-believers to do the same. Likewise, to be ethical is not merely to follow conventions. Just because something was done previously does not make it ethical. Kant and his followers widely recognized these points and responded with what he called "the autonomy of ethics." By this Kant argued that the Christian should not accept that something is right just because God commands it. Rather, explains Kant, God commands it because God recognizes that it is right.

What is distinctive and controversial about theories such as utilitarianism is their claim that the consequences are the only thing that matters. In contrast, Kantians put a high value on justice and commonly accuse consequentialist theorists of permitting, or even requiring, injustice. For example, Kantians argue that sometimes the only way to maximize happiness is to sacrifice the interest of some for the good of all, and that this is not acceptable. Returning to the example of punishment, retributivists claim that utilitarians would punish the innocent if the result were to promote the general welfare. So deontologists argue that utilitarians would favor deliberately faking evidence against an innocent person and putting him in jail if the result would deter future criminal activity.[6] The Australian philosopher H. J. McClosky suggests that a utilitarian sheriff in a preintegration southern town might feel forced to arrest and unjustly condemn an innocent black man for the unsolved rape of a white woman if that seemed to be the only way to avoid widespread violence by an unruly mob of white racists.[7] By a similar process, a utilitarian regulatory agency might seek to deter potential polluters by falsely accusing a firm of breaking clean air or water regulations if the results were to reduce environmental damage by less conscientious corporations. Deontologists would insist that whatever happens, the innocent do not deserve punishment. Indeed, they present a powerful argument that some things ought not be done even if they *would* produce great benefit.

Does utilitarianism sometimes require one to act unjustly? A utilitarian might reply that this criticism is superficial, for we must look beyond the immediate advantages of acting unjustly. Consider, for example, the effect on public confidence in the criminal justice system should knowledge of

---

6.   K. Armstrong. "The Retributionist Hits Back," *Mind*, vol. 70 (1961), pp. 471–490.
7.   H. J. McClosky. "A Non-Utilitarian Approach to Punishment," *Inquiry*, vol. 8 (1965), pp. 249–263.

unjust punishment become public. Such situations would have detrimental effects on the reputations of the officials involved. Utilitarians argue that, on balance, a utilitarian *would not* favor injustice.[8]

The issues here are of general relevance to environmental and other public policy decision-making. A person who believes, say, that factories should not be permitted to release *any* suspected carcinogenic substances into the environment, regardless of how, where, and in what quantities, would not be prepared to trade off the advantages of cheaper production against environmental impurity. Such a person would be acting on deontological principles. On the other hand, a person who is willing to accept some pollution in return for inexpensive consumer products is calculating the benefits and costs using utilitarian calculus.

Rule utilitarians defend themselves against the charge of sacrificing principles to expediency by claiming that rule-following itself has a positive utilitarian value. John Rawls and philosopher-lawyer Richard Wasserstrom argue that the existence of a clear, firm system of rules – such as the criminal law, or a generally accepted code of ethics – provides everyone with a certain security.[9] We can predict the behavior of our fellow citizens, and of institutions, because we can expect them to follow rules. Regardless of the content of the rules, it is valuable to know what they are and that they will be followed. In engineering, the knowledge that engineers will follow fundamentally sound rules reassures both other engineers and the public. Moreover, a system of rules permits and promotes impartiality and fairness since it tends to bureaucratize justice and ethics alike. Of course, the rules should also encourage efficiency and utility. But even in a case where breaking the rule would lead to more desirable consequences than following it would, one should obey the rule. Again, in the environmental area, rule utilitarians would not grant exceptions to regulatory standards merely to enhance immediate human well-being. To do so would forgo the advantages of a system of rules.

Rule utilitarians usually tie the obligation to follow rules to the system's being essentially beneficent. But what if the rules are generally bad ones and adherence to them would inhibit rather than promote happiness? Rawls believes that in general one ought to follow rules, while working through legitimate channels to change unjust rules. He also accepts that there may be

8.   T. L. S. Sprigge. "A Utilitarian Reply to McClosky," *Inquiry*, vol. 8 (1965), pp. 264–291.
9.   R. A. Wasserstrom. *The Judicial Decision* (Stanford University Press, Stanford, CA, 1961).

a right, or even a duty, of civil disobedience in the case of extremely unjust laws in a basically just system.[10]

A rule utilitarian engineer, then, might accept that rules of confidentiality should be followed, even when the general welfare might be advanced by breaking the rule. For instance, suppose you own a company that has developed a painkiller that is both extremely effective and safe. In the short run, the public might gain if you disclose the formula, thus allowing competition, reducing the price of the product, and benefiting the poor. But if companies cannot protect product information, they will have no incentive to invest resources in the development of new and better products.

Extreme cases might justify whistle-blowing, or breaking the confidentiality rule, as in the case of an irresponsible company intending to market some improperly tested or badly designed product. Rawls argues that the civil disobedient ought to accept the consequences, including arrest and penalty. Willingness to accept arrest and penalties shows that one is sincere and is acting on principle rather than for personal gain, thus distinguishing the civil disobedient from the criminal. Likewise, the engineer who publicly denounces a company's unethical conduct (and thereby becomes known as a troublemaker) plainly acts on principle, in contrast to the engineer who anonymously sells the story to a rival company or magazine for personal profit (and does not thereby jeopardize her/his professional career). As we note in Chapter 3, this self-sacrifice expects a lot of the engineer. But remember that for a utilitarian, each person's interests count equally. The utilitarian engineer is therefore committed to some degree of self-sacrifice if he/she is reasonably sure that blowing the whistle will prevent considerable harm to others.

Kant and some of his followers have argued that to be ethical is *necessarily* to follow a rule. In so doing they have asserted that ethical principles not merely rest on appeals to intuition but have a rational basis. As already noted, Kant maintained that one way to bring out the wrongness of a type of behavior is to show that it would not be adopted by a rational being. For instance, it is irrational to choose to cheat in business, or to fail to help others in distress because by cheating, the cheater adopts as a self-interested maxim: "Always cheat when I can get away with it." But can this be universalized – that is, can a rational person be prepared to accept it as applicable

10. J. A. Rawls. "The Justification of Moral Disobedience," in R. A. Wasserstrom, ed., *Today's Moral Problems* (Macmillan, New York, 1975).

to and by *all* persons? Since no rational being wants to be cheated (which would have to be the case, were the maxim universalized), the policy of cheating others turns out to be irrational. The same can be said of a policy of ignoring others who are in distress. What first appears to be in one's self-interest, when universalized, turns out to be opposed to it.

R. M. Hare, a contemporary Oxford philosopher (now at the University of Florida) in the Kantian tradition, argues that the distinctive feature of a moral principle (as opposed to a mere prejudice or a resolution to indulge oneself) is its general nature.[11] To call a thing good is to commend it to others. To assert that an action or principle is right is to prescribe it as a duty to yourself and to others. It would be quite irrational to assert, for example, that you had a right to pay kickbacks, and yet deny that anyone else had that right. To make use of ethical terms such as "right" you must commit yourself to the assertion of universal moral principles. To claim that you uniquely have the right to pay kickbacks to get public work is to place yourself in a specially favored category – which you cannot justify without giving special reasons. By giving reasons you are (whether you like it or not) asserting that *anyone in your position* has the same right. To claim the right to offer or accept bribes is therefore to assert paying kickbacks as a norm, which few would welcome.

Usefulness of Ethical Theories

Enough has been said so far, we hope, to show that it is possible for a reasonable person to adopt any of several approaches to ethics, each of which, however, has its deficiencies and problems. This does not mean that one cannot reasonably be, say, a utilitarian or Kantian deontologist, or ethical egoist. It does mean one has to accept that this ethical theory is not free from difficulties and may sometimes produce conclusions that are less than ideal. Ethical thinkers argue that their chosen theory is going to be right nearly all the time, and that it will create problems less often than alternative theories. Rather than assert that the theory is beyond criticism, ethicists continually strive to interpret, apply, and modify ethical theory to meet new objections and situations. In this respect, ethics is no different from any other human experience.

11. R. M. Hare, *The Language of Morals* (Oxford University Press, London, 1952).

Some critics claim that ethics is simply the dressing up of prejudices and special interests with arguments to make them appear respectable. But our ethical beliefs and arguments are not mere bias and cynical devices to defend self-interest. We may properly condemn attempts to defend what is wrong by devising theories and inventing facts to make one's actions appear right. But to have a clear conscience, and to defend our actions to others, we would like to believe that our actions are ethical. The desire to do the right thing and to have the approval of others is surely not itself unethical.

A disparity between thought and action creates uncomfortable feelings of what psychologists call "cognitive dissonance." The desire to reduce dissonance is legitimate, provided it is exercised honestly. Moreover, we want to defend our behavior as a rational response to how the world really is. We are most comfortable when we can justify our behavior by appealing to facts. Thus, we can defend our belief in the equality of men and women by showing that men and women, given equal opportunities, make equally good engineers. By supplying empirical evidence of this sort, we guard against prejudices. A sincere respect for the truth prevents ethics from degenerating into mere self-serving rationalization.

In this spirit, let us consider how one might go about justifying, sincerely and rationally, the institution of slavery. To a person socialized into a slave-owning society, it would be difficult to justify the owning of slaves and yet believe that one's slaves were even potentially one's equals in intellect or character. In sophisticated slave-owning societies, therefore, pseudoscientific theories arise to "justify" the institution of slavery. Thus Aristotle, who lived in a slave-owning society, tried to justify slavery on the premise that everyone is either a "natural" slave or master. He described slaves as typically of large stature and low intelligence, and masters as highly intelligent but unable to carry out heavy work. Aristotle believed that because the ability to reason and plan and the strength to carry out plans are not found in the same person, slavery is good for masters and slaves alike. He also believed that the master–slave distinction applied in barbarian (non-Greek) societies, but that relative to Greeks *all* barbarians could properly be regarded as slaves.[12]

A sincere believer in Aristotle's theory of slavery can feel no community with a slave, but instead can justify exploitation by appealing to the claim that slaves and masters need each other. Masters benefit most, but only they

12. Aristotle, *The Politics* I.3–I.7, trans. T. A. Sinclair (Penguin Books, Baltimore, 1962).

are rational. A slave is merely a useful living tool, like an ox but slightly more intelligent.

Aristotle's beliefs about slavery seem to many people to be self-serving, in that they appear to function as an apology for, or rationalization of, an oppressive, unjust institution. But we should not rush to accuse him of insincerity. Rather, we should apply the insights gained from studying Aristotle to our own ethical beliefs. Sexists, for instance, have traditionally viewed men as decisive, rational, calm, natural leaders and protectors; women as dithery, irrational, emotional beings in need of care and protection. Like Aristotle's theory of slavery, then, sexism comes with its own theory explaining the supposed inferiority of the exploited group.

In extremely inegalitarian societies, ethical questions concerning the relation of superior to inferior never arise. The exploiting class does not regard the exploited class as part of the moral community. In very sexist or racist societies, women or minorities are viewed as property to whom no ethical obligations are owed.

Moral progress, according to the ideals of Western society, begins when members of the dominant class start to see themselves as having duties toward the oppressed. This is the first step in the development of moral responsibilities. Extreme racists regard minorities as having no moral standing at all. More "enlightened" racists acknowledge that minorities have at least a degree of moral standing – for instance, the Victorian British imperialist view of indigenous people as entitled to the protection and sympathy of more "civilized" races because they were seen as unable to protect themselves from others or to run their own affairs. However, a true moral community cannot exist until all people view each other as equals, with a right to mutual respect and moral protection.

4.3   Ethical Theories Applied to the Environment

A parallel question about the scope of the moral community concerns human attitudes not only toward other human beings but also toward the rest of nature. Classical ethics has always been concerned exclusively with relations between humans. In the last century or so it has gradually been accepted that we have some obligations toward the non-human world. To cause gratuitous suffering to animals, for example, is generally thought to be wrong. This ethic is very selective, however, and does not extend to such "lower" animals as bugs or bacteria, let alone plants. Furthermore, we usually limit our obligations to the rest of nature to that which will serve humans. For example, we recognize our duty to preserve wilderness areas for aesthetic enjoyment and scientific study and to conserve resources for future generations, but we do not treat a tree with the same respect as we do a human.

Since at least the seventeenth century, the dominant secular stream of Western culture has sharply separated the human from the non-human by ascribing unique properties to humans. The *mechanistic* view of nature is associated with such philosophers as Galileo (1564–1642), Francis Bacon (1561–1626), Thomas Hobbes (1588–1679), and, in some interpretations, René Descartes (1596–1650). The living world is viewed as organized material and complex machines. Natural processes – including the behavior of animals – can be explicable causally. The human body is a machine, too: blood flows, food is digested, the senses receive impressions.

Some mechanists, such as Hobbes, regard humans as no more than machines. Others see humans as unique because they have the ability to reason. The Cartesian tradition (after Descartes), for example, credits humans with a non-material soul as well as a body, unlike other living things, which are merely bodies. Some Cartesians concluded that only humans can think, have emotions, make choices, or feel pleasure and pain. Since animals are mere machines, this argument continues, we cannot harm them any more than we might "hurt" a clock. Non-humans do not have interests and therefore do not deserve respect. Our treatment of animals, like our treatment of inanimate objects, is of no ethical significance. Seventeenth-century scientists carried out excruciatingly painful experiments on animals, secure in the knowledge that the animals did not suffer. (Their cries of pain were regarded as mechanically produced, like the squeak of an unlubricated wheel.) An echo of this mechanistic view of animals survives in the belief of some anglers that fish are unable to feel pain.

The Cartesian approach is not without support today. Contemporary philosopher Ruth Cigman, for instance, argues that animals cannot be said to have a right to life, since death is not a misfortune to them, and that even though animals desire to stay alive, it is not because of some long-range hope or ambition as to what they will do with their lives. In her view, killing animals is therefore not immoral, unless the act inflicts harm on humans.[13]

Most people today do not believe that animals are insensitive to pain, but

---

13. Ruth Cigman, "Death, Misfortune, and Species Inequality," *Philosophy and Public Affairs*, vol. 10, no. 1 (1981), p. 48.

they do believe that animals lack other mental characteristics. Many people condemn bullfighting, for example, but see nothing wrong with intensively raising animals for food, so long as no physical pain is caused. Nevertheless, such animals experience intense frustration, discomfort, deprivation, and boredom, but most people do not believe that such treatment of animals is immoral behavior.

Ignorance about animal behavior and ignorance about what happens to animals reinforce each other. Most people never see what goes on in intensive-rearing facilities, in slaughterhouses, and in the transport of livestock. They perceive the animal industry as humane, and therefore feel comfortable about eating meat. Likewise, they do not realize that new products (even nonessential ones such as hairspray) are routinely tested on animals and that millions of animals suffer and die every year in such testing. They therefore feel comfortable going out and buying new products. Finally, most people assume that medical research provides great benefits for humans at the cost of minimum, unavoidable suffering to animals.

The utilitarian approach to ethics recognizes suffering as a loss of happiness, and therefore an undesirable outcome. Some utilitarians have even extended this concept to the suffering of animals. Jeremy Bentham, one of the founders of utilitarianism, believed that the suffering of animals was an important consideration:

> The day may come, when the rest of the animal creation may acquire those rights which never could have been withholden from them but by the hand of tyranny. The French have already discovered that the blackness of the skin is not reason why a human being should be abandoned without redress to the caprice of a tormentor. It may come one day to be recognized that the number of the legs, the villosity of the skin, or the termination of the *os sacrum,* are reasons equally insufficient for abandoning a sensitive being to the same fate. What else is it that could trace the insuperable line? Is it the faculty of reason, or perhaps the faculty of discourse? But a full-blown horse or dog is beyond comparison a more rational, as well as more conversable animal, than an infant of a day, or a week, or even a month old. But suppose they were otherwise, what would it avail? The question is not, Can they *reason?* nor Can they *talk?* but Can they *suffer?*[14]

14. Jeremy Bentham, *Introduction to the Principles of Morals and Legislation* (1789), chap. 17, sec. 1, footnote to para. 4.

Some people believe that our treatment of animals does not cause suffering; that anticruelty and humane slaughter legislation has abolished animal suffering. These people do not recognize any animal interests other than avoiding physical pain. Just as Aristotle believed that slavery did not harm slaves, some people believe that in general we do not harm animals. When harm occurs, this is acceptable because important human interests – life and health – are at stake that might outweigh the interests of animals. Human interests come first because humans are so much *better* than animals. Humans alone have intellectual, moral, and spiritual capacities.

And yet, we recognize that we do owe some moral consideration to the non-human world. We can rightfully ask what our ethical theories can tell us about these attitudes, and how these theories can provide guidance for our actions toward nature.

Classical ethicists have developed two different approaches in order to respond to this question. The first approach is based on an instrumental value of nature. In this view, nature is only valuable to the extent that it benefits humans. The second approach is to argue that allowing humans to be cruel to animals will eventually cause them to be cruel to other humans as well.

The notion that nature is here only to benefit people is an old one. Aristotle argues:

> Hence it is similarly clear that we must suppose that plants exist on account of animals . . . and the other animals for the sake of man, the tame ones because of their usefulness and as food, and if not all the wild ones, then most, on account of food and other assistance [they provide, in the form of] clothing and other tools which come from them. If, then, nature does nothing without an end and nothing is in vain, it is necessary that nature made all these on account of men.[15]

Kant incorporated nature into his ethical theories by suggesting that our duties to animals are "indirect" duties, that is, duties only to humans. His view is quite clear: "So far as animals are concerned, we have no direct duties. Animals are not self-conscious and are there merely as a means to an end. That end is man."[16]

---

15. Aristotle, *The Politics*, trans. T. A. Sinclair.
16. Immanuel Kant, "Duties Toward Animals and Spirits," in *Lecture on Ethics* (1785), p. 240, as quoted by Mary Midgley, "Duties Concerning Islands," in R. Elliot and A. Gare, eds., *Environmental Philosophy* (Pennsylvania State University Press, State College, 1983).

This view of nature holds that since it is necessary to live in a healthy environment and to enjoy the other pleasures of life, it is rational that we be concerned with the environment. One could argue that to contaminate water or pollute air that other people enjoy is damaging something that belongs to others. It's a form of stealing. One could also argue that animals have utility and thus should be protected. We would not want to kill off all the plains buffalo, for example, because they are beautiful and interesting creatures, and we enjoy looking at them. To exterminate the buffalo would cause harm to other humans and this is the sole reason we should avoid exterminating the species. Philosopher William F. Baxter, for example, says that "damage to penguins, or to sugar pines, or geological marvels is . . . simply irrelevant. . . . Penguins are important [only] because people enjoy seeing them walk about the rocks."[17]

Not only would we not want to kill useful animals, but we would not want to exterminate species since they might somehow be useful in the future. An obscure plant or microbe might be essential in the future for medical research. We should not deprive others of that possible benefit. And we should not destroy the natural environment because so many people enjoy hiking in the woods or canoeing down rivers, and we should preserve these for *our* benefit.

The "instrumental value of nature" approach to environmental ethics has a number of limitations. First, it still allows us to kill or torture individual animals as long as this does not harm other people. Such an ethical mandate conflicts with our feelings about animals. We would surely condemn persons who caused unnecessary harm to any sentient animal. Second, the "instrumental value of nature" approach creates a chasm between humans and the rest of nature, a separation many people find unacceptable. Basing our environmental ethics only on instrumental values implies that the rest of the world is here only to serve human needs.

The second approach classical ethicists use to incorporate environmental attitudes into ethical theory concerns not human interests but human character. Philosophers such as Thomas Aquinas (1225–1274) and Locke argue that damaging the environment or causing pain to animals is immoral because such acts may harm our character. We should not allow people to cause pain in animals because such actions demean the perpetrator, and if people become accustomed to causing pain in animals, they may find it eas-

---

17. William F. Baxter, *People or Penguins: The Case for Optimal Pollution* (Columbia University Press, New York, 1974), p. 5.

ier to do the same to people. Locke wrote in 1693 that cruelty by children to animals should be corrected because it "will, by Degrees, harden their Minds even towards Men."[18] In this view, harm to animals is wrong only because it may lead to an increase in harm to people.

However, this approach is open to several objections. First, there is no evidence that cruelty to animals leads to cruelty to people.[19] Most studies indicate that children's cruelty toward animals results from other psychological problems, and although this may indeed be manifested later in life as cruelty toward people, it does not have its seeds in permitting cruelty toward animals. This does not seem to be a sound argument, therefore.

Today, most people believe that cruelty – in the sense of inflicting pain in order to enjoy the suffering of the victim or, more generally, inflicting pain without a very good reason – is wrong, regardless of victim or circumstance. Locke assumed a moral chasm between humans and the rest of the sentient world, such that our actions are acceptable as long as they do not cause us to become less ethical toward other humans. Today this is simply counterintuitive. Plainly, many animals can suffer pain, and as discussed in Chapter 5, we cannot dismiss pain just because the sufferer is non-human. As philosopher Mary Midgley has pointed out, "An ethical theory which, when consistently followed through, has iniquitous consequences, is a bad theory and must be changed."[20]

Another problem with Locke's argument occurs when we extrapolate it to include other parts of nature. The destruction of forests or natural habitats is wrong solely because of the effect these acts would have on humans, in that people who destroy natural things would be more likely to harm other humans. Again, this claim lacks empirical evidence. Do lumberjacks and butchers have a particular history of immorality? The destruction of nature might be an *effect* of a bad character, but not a *cause* of it.

4.4. Conclusions

Ethical theories are about how people ought to treat each other. We accept morality because it is in our self-interest to do so, and we expect others to do

---

18. John Locke, *Some Thoughts Concerning Education* (1693).
19. Eileen S. Whitlock and Stuart R. Westerlund, *Humane Education: An Overview* (National Association for the Advancement of Humane Education, Tulsa, OK), quoted in M. Fox, *The Case for Animal Experimentation* (University of California Press, Berkeley, 1986), p. 75.
20. Mary Midgley, "Duties Concerning Islands," in R. Elliot and A. Gare, eds., *Environmental Philosophy*, p. 168.

likewise. We understand that if we act morally, we will lead fuller lives and will be able to engage in personal as well as social endeavors without fear of being harmed. In other words, everyone benefits when we act ethically toward each other. We are all moral agents and a part of a moral community. Only members of this moral community deserve and receive moral protection and moral consideration.

Classical ethical theories, however, do not appear to respond to what our intuition tells us about the non-human world. We believe that we should care about non-human life and the natural environment, not for our selfish personal sake but because it seems to be the right thing to do. But there seems to be no ethical theory that will explain this attitude, nor does there seem to be a rational ethical argument for why everyone ought to behave in this way.

Perhaps the solution to this problem lies in expanding the moral community. If we can include animals and the environment generally in our moral community, perhaps we can learn how to treat them ethically. This possibility is the focus of the next chapter.

Suggested Supplemental Readings for Chapter 4

1. *Justification for the Use of Animals*, St. Thomas Aquinas
2. *Animals Are Machines*, René Descartes
3. *Why We Have No Obligations Toward Animals*, Immanuel Kant
4. *The Ethical Relationships Between Humans and Other Organisms*, R. D. Guthrie

5

The Search for Environmental Ethics by Extending the Moral Community

The moral community, as defined by traditional ethics, is the group of people to whom we owe moral consideration. In Aristotle's time, the moral community included only free men. With time, it became evident that the exclusion of slaves and women was both irrational and indefensible, and the moral community expanded.

With the inclusion of women and slaves in the moral community, it was not the criteria that changed but, rather, our perceptions of who meets the criteria. Aristotle did not accept women and slaves into his moral community, but he should have known better. It ought to have been clear to him, as it is to most of us, that almost all adult humans are capable of thought and language, can act in a reciprocal manner with other humans, and are therefore capable of functioning as moral agents.

A considerably stickier point has been what to do about nonrational or non-adult human beings who are *not* capable of becoming moral agents. We all agree that all humans should be included in the moral community, but this means that we cannot use the requirement of reciprocity as the criterion for inclusion.

Most ethicists recognize that reciprocity as a criterion for inclusion in the moral community does not work, and have responded in various ways. Their arguments claim that (1) if nonrational humans *could* express their feelings, they would choose to reciprocate; or (2) that we can place ourselves in similar circumstances and would in this case wish to be afforded moral protection and consideration. Other ethicists appear to consider the problem as marginal and unworthy of much effort.

Bernard Gert's theory of morality, for example, depends on the agents being both rational and impartial. He accepts only grudgingly that most people would insist on including infants and comatose people in the moral community. Such an acceptance, he argues, is emotional and is not because of any rational analysis. "In deciding who belongs in the group [moral community] toward which we should act impartially in accordance with the moral rules, the considerations that incline us one way or the other are often based on emotional considerations."[1] John Rawls acknowledges that permanently irrational human beings "may present a difficulty" but does not address the problem further.[2]

It is clear to most people that human beings of all kinds belong in the moral community, including the insane, the comatose, and, in some respects, even the dead. But historically, we have excluded all things non-human from the moral community because they fail to meet the reciprocity requirement.

As we conclude in Chapter 4, conventional ethics largely excludes animals and the rest of non-human nature from the moral community. If we wish to argue that some or all of the non-human world ought to have moral standing, we need to revise or reinterpret our criteria for admission to the moral community. In this chapter, we consider the possibility of developing an environmental ethic by extending the moral community while staying within the framework of Western ethics.

5.1   Extending the Moral Community

Many modern philosophers argue that the inclusion of non-humans within the moral community makes no sense whatever. According to Richard Watson: "To say an entity has rights makes sense only if that entity can fulfil reciprocal duties, i.e. can act as a moral agent." Watson goes on to argue that moral agency requires certain characteristics, such as self-consciousness, capability of acting, free will, and an understanding of moral principles. Since most animals do not fulfill any of these requirements, they cannot be moral agents and therefore cannot be members of the moral community.[3]

H. J. McClosky agrees, insisting that all humans are potentially moral agents, and "where there is no possibility of [morally autonomous] action, potentially or actually . . . and where the being is not a member of a kind

1.   Bernard Gert, *Morality* (Oxford University Press, New York, 1988), p. 89.
2.   John Rawls, *A Theory of Justice* (The Belknap Press of Harvard University Press, Cambridge, MA, 1971), p. 510.
3.   Richard A. Watson, "Self-Consciousness and the Rights of Nonhuman Animals and Nature," *Environmental Ethics*, vol. 1, no. 2 (1979), pp. 99–129.

which is normally capable of [such] action, we withhold talk of rights."[4] To these philosophers, the reciprocity principle limits the moral community to humans.

But have we not already disposed of reciprocity as the requirement for inclusion in the moral community? Perhaps some principle other than reciprocity should guide our understanding of the moral community. But once we crack open the door to non-humans, what are we going to let in? What can legitimately be included in our moral community, or, to put it more crassly, where should we draw the line?

Extending the Moral Community to Include Sentient Animals

One approach for extending the moral community is to assert that other creatures have rights similar to the rights enjoyed by humans. As discussed in Chapter 4, Locke and Hobbes argued that all humans have rights to life, liberty, and property, regardless of social status. These rights are natural rights in that we humans cannot give them to humans, or we would be giving rights to ourselves, which makes no sense. In this view, all humans are "endowed with inalienable rights" that do not emanate from any giver.

What if we suggest that non-human animals also have "inalienable rights" simply by virtue of their being, just as humans do? What if we claim that they have rights to exist, to live and prosper in their own environment and not to have humans deny them these rights unnecessarily or wantonly. If we agree that humans have rights to life, the right to liberty, and the right not to have pain inflicted upon them, then it seems only reasonable that animals, or at least those animals that can feel similar sensations, should have similar rights. And if other creatures have rights, they must also be included in the moral community. If we abandon the reciprocity requirement, we can admit non-humans into the moral community.

North Carolina State University philosopher Tom Regan in his well-known book *The Case for Animal Rights*[5] constructs his argument around a symmetry with human rights. First, he recognizes humans as having rights because we consider each human to have inherent value, that is, value in and of ourselves and not merely as a means to some external purpose. The reason why we value humans is that they are "subjects of a life": we perceive each human not just as a biological entity but as a *person*, an entity that has

---

4.   H. J. McClosky, *Ecological Ethics and Politics* (Rowman & Littlefield, Totowa, NJ, 1983), p. 29.
5.   Tom Regan, *The Case for Animal Rights* (University of California Press, Berkeley, CA, 1983).

experiences and capacities, hopes and fears, with its own life to live. If that is indeed why we value humans and accord them rights, then to be consistent we should also value similarly non-human beings that possess similar qualities. Because many animals, especially adult mammals, do possess these qualities, we ought to include them in the moral community and cease to kill them for food, use them as subjects for medical and other research or product testing, or exploit them in any other way for our benefit. Given the status quo, Regan argues, we are thoroughly unjust in our treatment of animals. We accord moral equality to humans who fail to meet the criteria for valuing humans – for instance, severely and irreversibly brain damaged infants – whose lives are no more than a series of simple biological events. Yet we treat intelligent, curious, social animals as mere research subjects, kill them for various products, and exterminate them by polluting or destroying their habitats, none of which we would be prepared even to consider in the case of the most marginal of humans.

In agreeing with Bentham and Regan, Eric Mathews argues that since we humans extend moral concern toward those humans who cannot reciprocate, we should also include in this circle animals that can suffer. He concludes, "There is no logical reason for distinguishing in this regard between human beings and other sentient creatures, so that the grounds for opposing cruelty to animals are precisely the same as those for opposing cruelty to our fellow human beings."[6]

An argument based on rights has a serious problem, however. If we accept that at least some animals have rights, then we must decide whether to agree that the rights of animals are equal to those of humans, or somehow to list and rank the animals and the rights in order to specify which rights animals have under what circumstances. In the first instance, killing your neighbor and squashing a cockroach would have equal moral significance. In the second, we would have to decide that the life of a chickenhawk is less important than the life of a chicken, and make an infinite number of other comparisons.

R. D. Guthrie, for example, argues that "a sliding scale of morally obligatory conduct toward, or expected from, other organisms . . . is operationally unfeasible. We would have to formulate some sort of discriminatory system which would define the nature and extent of preferential moral treatment," a task that Guthrie considers impossible.[7]

6.   Eric Mathews, "The Metaphysics of Environmentalism," in N. Dower, ed., *Ethics and Environmental Responsibility* (Avesbury, Aldershot, UK), p. 39.
7.   R. D. Guthrie, "The Ethical Relationship between Humans and Other Organisms," *Perspectives in Biology and Medicine,* vol. 11 (1967), pp. 52–62.

Most writers who consider sentience to be the moral cutoff do so without arguing for animal rights. Philosopher Mary Anne Warren, for instance, suggests that the suffering of animals is an evil of the same sort as the suffering of humans. For that reason, inflicting pain on any creature is wrong and immoral. She argues that the "essential reason for regarding torture as wrong is that it *hurts,* and that people greatly prefer to avoid such pain – as do animals."[8] Warren believes that arguments asserting that animals should not to be tortured because they have rights will not be convincing.

Australian philosopher Peter Singer uses utilitarian principles to extend the moral community to include sentient animals. In his book *Animal Liberation* he points out that the discrimination against women and less advantaged races is condemned as sexism and racism because it systematically ignores the interests of those groups in favor of the interests of the dominant group.[9] Most of us agree that we have an obligation to give equal weight to equal interests, regardless of the race or sex of a person. Singer argues that the same is true of species membership: a being's interests matter regardless of the species to which the individual belongs. Utilitarianism, the ethical theory to which Singer subscribes, acknowledges the equal interest that all sentient beings have in avoiding suffering and the wrongfulness of inflicting suffering on any sentient beings without sufficient justification. Singer therefore condemns our exploitation of animals for food and the like as "speciesism," comparable to racism and sexism.

An environmental ethic based on sentience faces various problems, however. First, we do not know for sure what animals feel pain. We can assume that higher animals feel pain because their reactions to pain resemble ours. A dog howls, and a cat screams and tries to escape the source of the pain. But what about creatures that cannot show us in unambiguous ways that they are feeling pain? Does a nightcrawler feel pain when I put it on my fishhook? Does a butterfly feel pain when I stick a pin through its body? Of course, the impossibility of drawing a precise line does not necessarily invalidate the theory; even if we decide to stop inflicting pain on only those animals clearly capable of suffering, we avoid inflicting a great deal of pain. But there would still be a considerable area of uncertainty.

If the utilitarian approach is used, the *amount* of pain suffered by animals and humans must be calculated. If, for example, a human needs an animal's

8.   Many Anne Warren, "The Rights of the Nonhuman World," in R. Elliot and A. Gare, eds., *Environmental Philosophy* (Pennsylvania State University Press, State College, 1983), p. 114.
9.   Peter Singer, "Animal Liberation: Ethics for Our Treatment of Animals," *New York Times Book Review,* 5 April 1973.

fur to keep warm, is it acceptable to cause suffering in the animal in order to prevent suffering in the human? Cost-benefit calculations involving only humans are difficult enough; how can we quantify animals' pain in order to figure them into a decision-making procedure?

By focusing on pain, the utilitarian approach also ignores the problem of instant, painless death. As Michael Fox points out, "If the death of an animal (or indeed a human) could be brought about without accompanying suffering, what, on utilitarian grounds, could be judged morally wrong with such an act?"[10]

The issue of sentience becomes even more complicated when we consider not just animals but also plants. Many people consider it wrong to destroy at least some plants – a magnificent tree that stands in the path of a new sewer line, for example, as in Case 3 in Chapter 1 – but the sentience criterion does not provide any reason to include plants in the moral community. Certainly, some people insist that plants feel pain and that we are just too insensitive to recognize it, but so far the evidence to support this claim is unconvincing.

Further, neither the rights nor the utilitarian theories can account for the value we attribute to members of endangered species. Indeed, both Singer and Regan emphasize that because species membership is irrelevant to moral status, an individual of a rare species is no more valuable than a member of a common species. Regan writes,

> The rights view is a view about the moral rights of individuals. Species are not individuals, and the rights view does not recognize the moral rights of species to anything.[11]

The sentience criterion, whether based on rights or utilitarian theory, does not make a distinction between species. It is, accordingly, no worse to harm a member of a rare species than a member of a common species, other things being equal.

Finally, a problem with an environmental ethic built purely on sentience is that such an ethic does not provide any reason to preserve natural environments, except when the destruction of these environments harm sentient creatures. In adopting the sentient animal criterion as our environmental ethic, we would face no moral obstacles against, for example, destroying the Grand Canyon.

---

10. M. Fox, *The Case for Animal Experimentation* (University of California Press, Berkeley, 1983), p. 66.
11. Tom Regan, *The Case for Animal Rights*, p. 359.

In summary, both utilitarianism and rights theories include at least some non-human animals within the moral community. To the extent that environmental concern is concern about the lives and suffering of animals, such extensionism enables us to give a coherent account of why it is wrong to carry out various environmentally damaging acts without abandoning conventional ethical theories. However, extensionist ethics based on sentience cannot provide reasons for preserving and protecting endangered species, or plants, or natural features. For many people an ethic based on sentience seems to miss the point in failing to explain why is it wrong to exterminate a whooping crane, for example, or to destroy a unique rock formation.

Can we eliminate some of these criticisms by drawing the line between the "ins" and the "outs" somewhere else?

Extending the Moral Community to Include All Life

Logically, the next extensionist step is to include all of life in the moral community. This is not as outrageous as it might seem, the idea having been developed by Albert Schweitzer (1875–1965), who called his ethic a "reverence for life." He argued that limiting ethics to only human interactions is a mistake, and that a person is ethical "only when life, as such, is sacred to him, that of plants and animals as that of his fellow men."[12] Schweitzer believed that an ethical person would not maliciously harm anything that grows, but would exist in harmony with nature. He recognized, of course, that in order to eat, humans must kill other organisms, but he held that this should be done with compassion and a sense of sacredness toward all of life. To Schweitzer, humans beings are simply a part of the natural system.

Charles Darwin is probably most responsible for the acceptance of the notion of humans as not being different in kind from the rest of nature. Since humans evolved from less complex animals, we are different only in degree and not in kind from the rest of the animals. We are simply another part of a long chain of evolution. As Australian environmental philosopher Janna Thompson points out: "Evolutionary theory, properly understood, does not place us at the pinnacle of the development of life on earth. Our species is one product of evolution among many others."[13]

12. Albert Schweitzer, *Out of My Life and Thought: An Autobiography* (New York, 1933).
13. Janna Thompson, "Preservation of Wilderness and the Good Life," in R. E. Elliot and A. Gere, eds., *Environmental Philosophy* (Pennsylvania State University Press, State College, 1983), p. 97.

Similarly, City University of New York philosopher Paul Taylor holds that all living organisms have *inherent worth* and therefore are a part of the moral community. To deny them membership is to treat them unjustly. What he calls the "biocentric" outlook depends on the recognition of common membership of all living things in the earth's community. Each organism is a center of life, and all organisms are interconnected. For Taylor, humans are no more or less important than other organisms.[14]

Both the Schweitzer "reverence for life" and the Taylor "biocentric" philosophies bring all of life into the moral community, including plants. The notion of plants having rights was first proposed by Christopher Stone in his extremely influential paper "Should Trees Have Standing?"[15] Stone argues that if we yield rights to inanimate entities such as corporations, why then should natural objects such as trees not also possess legal rights? By virtue of standing (rights), trees (and other plants) deserve an equal place in our moral community.

This approach to environmental ethics has considerable appeal, and numerous proponents. However, it suffers problems similar to those raised against the sentience approach. First, there is no agreement about where to draw the line between the living and non-living; indeed, there may be no such sharp distinction. Viruses present the greatest problem here since they seem to possess characteristics of both living and non-living things.

But what if we do decide that viruses are living organisms? Taylor's view would then require that the polio virus and other organisms that cause disease and death in humans – and in other animals – also be included in the moral community.

Once again, we must decide how to weigh the value of non-humans relative to the value of humans. Should the life of all creatures be of equal value, so that the killing of a pine tree is just as wrong as the murder of a human being? Advocates of this view presumably want to promote the moral status of whales but the view is more likely to result in a greatly reduced moral status for humans.

If the claim of equality is implausible, there will have to be some scale of values, some ranking in the value of living organisms. But on what basis? Are microorganisms of equal value to polar bears? Should we hold a rattlesnake

---

14. Paul W. Taylor, *Respect for Nature: A Theory of Environmental Ethics* (Princeton University Press, Princeton, NJ, 1986).

15. Christopher Stone, *Should Trees Have Standing?* (William Kaufmann, Los Altos, CA, 1974).

in the same reverence as the humpback whale? Again we must draw a line somewhere between those creatures to be included in the moral community, and those that are excluded. "You, the amoeba, you're in. You, the parame-cium, you're out" just does not compute.

Such ranking also introduces difficulties in determining what practices are morally acceptable. Is it, for example, better to eat cows grown on the range than chickens grown in a chicken "farm"? Does acknowledging the value of life require vegetarianism? Even if everybody becomes a vegetarian, we would have to value some plants more than others. Crop plants might be ranked higher than the wild plants and animals that have to be cleared off to make room for farms and gardens. In terms of respect for life, is it better to use an electric car and promote the production of nuclear waste, or drive a gasoline-powered car and produce carbon dioxide, a global-warming gas?

Finally, it is not so clear that we value life for its own sake, even human life. Being (just) alive is not necessarily better than being (just) dead, and maybe worse if one is suffering terrible pain and is terminally ill. Life seems, rather, to be a *condition* of having valuable experiences and activities; and death is a tragedy because it means the end of experiences and activities. The life of an animal or a plant, likewise, is valuable because it is a condition of the animal or plant flourishing in some way, though in the case of plants, flourishing may be very different from that of humans.

We must conclude that extending the moral community to include all life does not result in a useful environmental ethic.

Extending the Moral Community to Include Systems

If we cannot value all living things equally, we might have more success if we place value on the ecosystems within which we all live. When we value sys-tems instead of (or as well as) individuals, we simply recognize that ecosys-tems are essential for the growth and development of *all* organisms. If the ecosystem is damaged, the organism must either adapt or die. Thus, the value is actually in the system and not in the individuals because the best way to protect all individuals is the protect the system.

The most important proponent of this line of reasoning is Aldo Leopold (1887–1947), who wrote that "the individual is a member of a community of interdependent parts," and that removing one part may cause the entire sys-tem to collapse. He argued that a new ethic should change the role of people from being the conquerors of nature to simply being members of the larger "biotic community."[16] Leopold's holistic ethic "enlarges the boundaries of the community to include soils, water, plant, and animals, or collectively, the land." This changes the role of humans from masters of the planet to simply members of the biotic community. Leopold's most concise statement of the land ethic is often quoted:

> A thing is right when it tends to preserve the integrity, stability and beauty of the biotic community. It is wrong when it tends otherwise.

Although Leopold described the land ethic as an extension of classical ethics, some philosophers have suggested that the land ethic is new and radically different. To John Rodman, because the land ethic recognizes value in systems and connections rather than in individuals it represents a paradigm shift.[17]

Leopold's land ethic is highly attractive for a number of reasons. First, it is nonanthropocentric, placing the human not at the pinnacle but within a larger community. The ethic also supports many of our attitudes about the environment. For example, we no longer need to be concerned with killing insects so long as we preserve the ecosystem in which we, and the bugs, function. One can, in fact, extend this idea and suggest that the system determines the value of the individual or species, and not the other way around, and therefore it makes sense that the system is of greatest ethical concern. J. Baird Callicott calls this the "ecocentric" approach to environmental ethics.[18]

Concentration on the health of supporting systems instead of individual organisms clarifies our obligation to future generations. The cutting of a tree or the killing of an animal in isolation has no bearing on future generations, unless they are integral parts of the system that will be necessary to support future humans. The destruction of the rain forests is wrong because if these valuable ecosystems are irreparably destroyed, the global ecology will change. Deforestation may cause, along with the predicted greenhouse effect, desertification to occur, or in the extreme, could even make the earth uninhabitable in the not too distant future. Nevertheless, while we understand what effect these actions could have, there will be no compelling rea-

16. A. Leopold, *A Sand County Almanac* (Ballantine, New York, 1966; originally published in 1949).
17. J. Rodman, "Four Forms of Ecological Consciousness Reconsidered," in D. Scherer and T. Attis, *Ethics and the Environment* (Prentice-Hall, Englewood Cliffs, NJ, 1983), pp. 82–92.
18. J. B. Callicott, *In Defense of the Land Ethic* (SUNY Press, Albany, NY, 1989).

son to do anything to prevent this from occurring unless our ethic values the preservation of ecosystems.

A possible unwelcome feature of this approach to environmental ethics is that if the system is important, the individual is correspondingly less so, and it would be perfectly appropriate to kill certain organisms for the sport of it, knowing that the system will recover. This ethic also does not address the question of cruelty; it is concerned only with the presence or absence of organisms in the system. Taken to its logical conclusion, it would be perfectly acceptable to thin out the human population as well since this species is causing by far the greatest harm to the global ecosystem. Understandably, such notions are repugnant to most people. Tom Regan, for example, has called the ecocentric approach to environmental ethics "environmental fascism."[19]

Another concern about the land ethic, from the ecological perspective, is that any given ecosystem is seldom stable, even in the climax stage. Some organisms, for instance, function only during severe perturbations within ecosystems, such as the months or years following a forest fire or flood. What is the "stability" of an ecosystem that has been severely damaged, and if this is a natural phenomenon, is it therefore unethical to prevent such natural perturbations as forest fires? If a natural fire occurs in a forest, should humans fight it, or let it burn freely?

Approaches like Leopold's have their greatest appeal to environmentalists who value ecosystems and processes and who believe that environmental damage has been caused mainly by our anthropocentric way of looking at the world. They believe that concern with individual animals or other living things diminishes the essential fact that the world operates as a system. But as we note above, if thinking about individuals tends to downgrade or ignore whole systems, thinking holistically conversely reduces or ignores the significance of individuals.

This difficulty is not unique to environmental thinking. Social thought and ethics requires us to balance concern for individuals with concerns for groups. A similar balancing might be possible for environmental issues.

The most important criticism with the ecocentric approach to environmental ethics is that we cannot answer the question "Why *should* I value the ecosystem?" If we accept the earth as one large ecosystem, for example, everything falls into place. We are each a part of a single organism with a responsibility to not cause harm to that larger organism (the earth), or else

---

19. Tom Regan, *The Case for Animal Rights*, pp. 361–362.

either the earth will die or people will be wiped out like some pathogenic organism. But there is no compelling reason to accept this hypothesis, just as there is no compelling reason to bring smaller ecosystems such as lakes, rivers, or rain forests into our moral community.

Extending the Moral Community to Include Everything

One means of removing the objection of knowing where to draw the line is to simply extend the line to include everything within the circle of moral concern. Everything – plants, humans, animals, things – becomes intrinsically valuable and is thus a legitimate object of our moral concern. These objects in nature are of value in and of themselves, and not because of any value placed on them by people. This view, argued by Val and Richard Routley[20] and Homes Rolston III,[21] leads to the conclusion that everything, including humans, should be valued equally.

Support for this position within the Western tradition comes from the seventeenth century, when Baruch Spinoza (1632–1677) argued that, contrary to the theories of René Descartes, every being is part of a larger whole and part of an ever-changing universe. Spinoza recognized that when people die their bodies become parts of other objects such as plants, thus demonstrating an understanding of ecology. Since everything has value in God's eyes, he argued, everything has an equal right to exist.[22]

Tom Regan presents a similar concept as a "preservation principle," a principle of "non-destruction, interference, and generally, non-meddling."[23] By including everything in our moral community, he eliminates the problem of where to draw the line between those things that are in and those things that are out. But this does not solve the problem of hierarchy. As Gene Spitler points out, within this ethic we would "find that shooting [our] neighbor was no more morally reprehensible than swatting a fly or stepping

---

20. Val Routley and Richard Routley, "Against the Inevitability of Human Chauvinism," in Kenneth Goodpaster and Kenneth Sayre, eds. *Ethics and the Problems of the 21st Century* (University of Notre Dame Press, Notre Dame, IN, 1987).
21. Homes Rolston III, *Environmental Ethics* (Temple University Press, Philadelphia, PA, 1988).
22. George Sessions, "Western Process Metaphysics: Heraclitus, Whitehead, and Spinoza," in Bill Devall and George Sessions, eds., *Deep Ecology* (Gibbs Smith, Salt Lake City, UT, 1985).
23. Tom Regan, "The Nature and Possibility of an Environmental Ethic," *Environmental Ethics*, vol. 3, no. 1 (1981), pp. 31–32.

on a wildflower."[24] But if we assume that humans *are* more valuable than flies, we must describe those characteristics that make them more valuable. As soon as we start to propose some hierarchical characteristics, we are back to the original problem of judging everything by human standards.

Another, perhaps more subtle, criticism of Regan's view of nonmeddling by humans is that this ethic would require humans to act in ways different from what they would normally do. That is, humans are by their nature exploitative, and have been using the rest of nature for their benefit for as long as there have been humans. How is it then possible suddenly to convince humans that they should no longer change the ecology of the planet? If the natural behavior of humans is to exploit, then aren't we already doing "what comes naturally?"[25]

Recall Case 10 in Chapter 1 about Carole, who would not run over a box turtle. Carole may have decided to not run over the turtle because she felt it had value. But it had to have received that value by one of two ways: she had to have assigned it this value, or the turtle had an intrinsic value of and by itself, irrespective of her evaluation of it.

If the first case is true, and if we indeed placed a value on the turtle, why did we do that? The "everything has a value" approach cannot answer this question. And if the turtle had a value all by itself, how did Carole know that it did? It did not have a sign on that read "VALUABLE TURTLE."

5.2 Conclusions

We argue that reciprocity should no longer be a necessary criterion for admission to the moral community, and therefore it is theoretically possible to admit non-human creatures, plants, places, systems, and so on.

The difficult decision, of course, is where to draw the line between those creatures and things that are to be included in the moral community and those that are not. If we choose not to draw a line, we must establish a hierarchy in which the rights of humans, non-human creatures, and inanimate things must be compared. Either we have not to draw a line, and extend the moral community to include everything and value everything equally, or we

---

24. Gene Spitler, "Justifying a Respect for Nature," *Environmental Ethics,* vol. 4, no. 3 (1982), pp. 255–260.
25. P. A. Vesilind, comment on Homes Rolston III, "Can and Ought We to Follow Nature?" *Environmental Ethics,* vol. 1, no. 4 (1979), p. 379.

have to draw the line somewhere and establish a hierarchy of how much we value each organism, system, or thing. This hierarchy must be created by humans, of course, and so it will no doubt be anthropocentric.

Neither option seems reasonable, and we are left with the conclusion that the environmental ethic cannot be developed by the technique of expanding the moral community. All expansionist ethical theories therefore have a fatal flaw, and we cannot rely on them to provide us with the comprehensive environmental ethic we seek.

This conclusion does not mean that extensionist approaches to environmental ethics are without value. Singer's arguments for caring for animals that feel pain, Schweitzer's reverence for life, and Leopold's land ethic – all have merit. Indeed, using any of these theories for making decisions concerning the environment is far better than saying that because there is no theory that works all the time, it is perfectly acceptable to not care about the non-human world. We certainly do not suggest such an alternative. We merely argue that whatever extensionist ethic one wishes to accept as the most plausible, one cannot demonstrate its validity to someone who starts from different assumptions. But that's not important. What is important is that one understands that these arguments are useful stepping-stones in the eventual development of a comprehensive environmental ethic.

Suggested Supplemental Readings for Chapter 5

1. *The Case for Animal Rights,* Tom Regan
2. *The Land Ethic,* Aldo Leopold
3. *Reverence for Life,* Albert Schweitzer
4. *Respect for Nature,* Paul Taylor
5. *Should Trees Have Standing?,* Christopher Stone
6. *Sierra Club vs. Morton, Secretary of the Interior*
7. *The Rights of Natural Objects,* Kristin Schrader-Frechette

6

The Search for Environmental Ethics in Spirituality

It should be evident from the preceding chapters that neither classical ethical theory nor extensionist theory leads to a comprehensive environmental ethic. Neither approach justifies in a rational manner our inner feelings of concern for individual organisms, places, ecosystems, or future generations.

Most environmental ethicists come from the mainstream empiricist-analytical philosophical tradition, and these writers use exactly the same terminology and reason in exactly the same way as in traditional ethics. But such mainstream approaches cannot help us develop a comprehensive environmental ethic. If we are to get anywhere, we clearly have to "think out of the box" and search for an environmental ethic from a different perspective.

The alternative perspectives discussed in this chapter all reject what we call the mainstream tradition and invite us to adopt a different way of looking at the world. For convenience, we refer to these alternatives as "spiritual," without meaning to imply that they are necessarily religious in the conventional sense. Because advocates of alternative paradigms reject what they see as the excessive emphasis on rationality and scientific method in the mainstream, their views may initially not appeal to engineers, who may think them unworthy of consideration because they lack a provable foundation. Certainly, it is not possible to "prove" assumptions of the spiritual approaches to environmental ethics, but neither is it possible to prove the foundations of engineering, which depend on unprovable assumptions about Newtonian mechanics, Euclidian geometry, and the truth of sense perception. We tend to take these assumptions for granted, to assume that we *know* that this is *really* how the world is, yet anyone who has studied geometry or logic knows these systems cannot proceed without making unprovable assumptions. The perspectives discussed in this chapter are simply more up front about the status of their foundations than we are in our everyday and professional lives.

From this perspective, we create an ethic much as we create a religion. That is, we begin with an unprovable belief and then use reason to deduce how we should behave. Suppose, for instance, we asked you to believe there is only one god, and that he/she has told us everything there is to tell in a book. If you are willing to believe that this is true, then we can build a religion based on this faith. Or, suppose we asked you to believe that a certain person by virtue of her/his position or birth is infallible in matters of religion. From this assumption, we develop a complex religion and use reason to determine our values and practices. Nevertheless, all of these rational deductions depend on that first assumption.

How does such spirituality, expressed in the form of a religion, develop? John Passmore observed:

> A morality, a religion, is not . . . the sort of thing one can simply conjure up. It can only grow out of existing attitudes of mind, as an extension or development of them, just because, unlike a speculative hypothesis, it is pointless unless it actually govern's man's conduct.[1]

It is quite obvious that a meaningful spirituality concerning the environment does not have to be "conjured up." All we have to do is look around and see how millions of people are restructuring their lives and making significant decisions based on what they perceive to be good for the environment. Such a spirituality must rise from a deep sense of justice and caring about the world and its inhabitants. Why not use this spirituality as a starting point for the development of an environmental ethic? While the classical philosophies depend on the Enlightenment tradition of reason and empiricism, it might be that such arguments lead to dead ends, and that spirituality is the clearest explanation for our attitudes toward nature. How else can we explain, for example, why some people "avoid making unnecessary noise in the forest, out of respect for the forest and its nonhuman inhabitants,"[2] if it cannot be explained on the basis of spiritual attitudes?

In this chapter we discuss several spiritual approaches to environmental ethics – transcendentalism, deep ecology, ecofeminism, the Gaia hypothesis, the Judeo–Christian tradition, Asian religions, Islam, animism, and modern pantheism. In each case the environmental ethics appeal to a faith

---

1.   John Passmore, *Man's Responsibility for Nature* (Scribner, New York, 1974), p. 111.
2.   Richard and Val Routley, "Human Chauvinism and Environmental Ethics," in D. Mannison, M. McRobbie, and R. Routley, *Environmental Philosophy* (Research School of Social Sciences, Australian National University, Canberra, 1980), p. 130.

that ultimately is not based on rational argument but affords an opportunity to construct a useful and satisfying environmental ethic.

6.1 Transcendentalism

One of the earliest American attempts to incorporate the spirituality of nature into a religion is transcendentalism. Two of the most widely read figures in this movement, Ralph Waldo Emerson (1803–1882) and Henry David Thoreau (1817–1862), rejected scientific empiricism and rationality as a means of understanding nature and instead substituted a human reality that transcends such worldly experience. The search for this deeper understanding of nature was the core of the transcendentalist movement that flourished during the mid-nineteenth century. Since wilderness was unspoiled by human activity, it was there that humans could come into contact with their deeper reality. To the transcendentalists, wilderness became sacred because this was where spirituality would be found.

Emerson believed that the universe consisted of the Soul and everything else, which he called Nature (capital N). By yielding oneself to Nature, humans can attain true spirituality. Consider an example of Emerson's florid prose:

> Standing on the bare ground – my head bathed by the blithe air, and uplifted into infinite space – all mean egotism vanishes. I become a transparent eyeball; I am nothing; I see all; the currents of the universal Being circulate through me; I am part or parcel of God.[3]

Transcendentalists learn the truth about human life by looking at the universe. Only by contemplating nature can we understand the spirit that provides life. Transcendentalists do not distinguish the sacred from the profane. Everything is sacred.

Transcendentalism in the nineteenth century was not a formal religion but a set of ideas held among a loosely knit group of intellectuals in New England. Its most prominent writers, especially Thoreau and Emerson, greatly influenced the conservation and preservation movements. John Muir, for example, used many of Emerson's allusions to nature in his arguments for the preservation of the Yellowstone Valley, and Muir's writings in

---

3. Ralph Waldo Emerson, "Nature," in *The Best of Ralph Waldo Emerson* (Walter J. Black, New York, 1941), p. 76.

turn had a great effect on President Teddy Roosevelt's decision to preserve the park (an example of where environmental ethics actually made a difference in the political process).

Perhaps because it was never institutionalized, the transcendental movement eventually died out. Its ideas, however, remain fresh, especially in the writings of Thoreau, who was an unabashed supporter of the usefulness of wilderness. To Thoreau,

> Life consists of wilderness. The most alive is the wildest. Not yet subjected to man, its presence refreshed him. One who pressed forward incessantly and never rested from his labors, who grew fast and made infinite demands on life, would always find himself in a new country or wilderness, and surrounded by the raw material of life. He would be climbing over the prostrate stems of primitive forest-trees. Hope and the future for me are not in lawns and cultivated fields, not in town and cities, but in the impervious and quaking swamps.[4]

But such romantic notions of wilderness do not make for a robust environmental ethic. An environmental ethic founded on an appreciation for wilderness results in a problem common to all spiritual approaches. We might appreciate wilderness because of the spiritual sustenance it gives us, but we can only find such sustenance if we appreciate the wilderness. If you see wilderness as dangerous and ugly, or locate the soul of humanity in the bustle of the city, who is to say that your urban vision is inferior to Thoreau's bucolic romanticism? An appreciation of wilderness has to originate not in reason but in some spiritual sensitivity.

6.2   Deep Ecology

The deep ecology movement is in some ways an extension of transcendentalism, and deep ecologists often cite transcendentalists such as Muir and Emerson. This movement was formalized with the writings of Norwegian philosopher Arne Naess (1912–), who proposes that environmental ethics should evolve from two fundamental values – *self-realization* and *biocentric equality*. These values are intuitive and defy rational justification, thus placing deep ecology firmly in the realm of spirituality.

---

4. Henry David Thoreau, "Walking," in *The Writings of Henry David Thoreau*, vol. 5 (Houghton Mifflin, Boston, 1906), p. 226.

Self-realization is the recognition of oneself as a member of the greater universe and not just as a single individual or even as a member of a restricted community. This self-realization can be achieved, according to Naess, by reflection and contemplation. The tenets of deep ecology are not inherently tied to any particular religious belief, though like Marxism or any other set of beliefs and values, it may function as a religion or worldview for the believer. As described by Kirkpatrick Sale:

> Most people in deep ecology have had the feeling – usually, but not always in nature – that they are connected with something greater than their ego. . . . Insofar as these deep feelings are religious, deep ecology has a religious component . . . [a] fundamental intuition that everyone must cultivate if he or she is to have a life based on value and not function like a computer.[5]

The second tenet of deep ecology is biocentric equality. According to Naess, this follows from self-realization, in that once we understand ourselves as one with other creatures and places in the world, we cannot regard ourselves as superior. Everything has an equal right to flourish, and humans have no special rights. We must eat and use other creatures in nature to survive, of course, but we must not exceed the limits of our "vital" needs. To a deep ecologist, the collecting of material wealth, or goods above the vital necessities, is unethical.

Someone who accepts these propositions is probably committed to the following corollaries:[6]

1. All life, human and non-human, has value in itself, independent of purpose, and humans have no right to reduce its richness and diversity except for vital needs.
2. Humans at present are far too numerous and intrusive with respect to other forms and the living earth, with disastrous consequences for all, and must achieve a "substantial decrease" in population to permit the flourishing of both human and non-human life.
3. To achieve this requisite balance, significant change in human economic, technical, and ideological structures must be made, stressing not bigness, growth and higher standards of living, but sustainable societies emphasizing the (non-material) quality of life.

5. K. Sale, "Deep Ecology and Its Critics," *The Nation,* 14 May 1988, pp. 670–675.
6. Ibid.

To deep ecologists, wilderness has a special value, and must be preserved not only for the psychological benefit of humans but, more importantly, out of respect for its intrinsic value as providing a place for other species to survive unhindered by human activity. Deep ecologists seek a sense of place in their lives, a part of the earth that they can call home. Deep ecologists also oppose the capitalist/industrial system that creates what they view as unnecessary and detrimental wealth. They disdain the idea of stewardship since it implies human decision-making and human intervention in the workings of natural environments. They seek to regain the sympathetic and supportive relationship between people and nature held by ancient peoples.

Deep ecology, with its emphasis on self-realization, can be intensely personal. In fact, some philosophers criticize deep ecology precisely because of what they consider to be its excessive focus on the spiritual development of the individual. But most tenets of deep ecology require large-scale changes involving the whole population. The precepts of deep ecology challenge our fundamental ways of doing business. Critics sometimes characterize it as non-humanist, because it does not give humans special benefits, and indeed indicts humans as the exploiters of nature. Especially galling to critics is the deep ecologist's call to depopulate the world, an objective that many deride as fascist and dehumanizing.[7] There is no doubt that if we are to have a less adverse effect on nature, we must first reduce the human population by a significant portion (90 percent has been suggested by some deep ecologists). Critics also argue that the acceptance of deep ecology as a global philosophy would revert civilization to the hunter-gatherer societies.

Finally, some critics of deep ecology argue that it is unacceptably elitist: that while its vision of a world largely reverted to wilderness and with far fewer people may appeal to a small number of the relatively rich and highly educated, it has nothing to offer to the great masses of poor humans trying to survive.[8] And because deep ecology requires voluntary acceptance of a lower material quality of life, it is unlikely that many people, even among the wealthy, will voluntarily accept its basic tenets of self-realization and biocentric equality. Nevertheless, deep ecology offers an attractive approach to environmental ethics once the basic tenets are intuitively accepted. Even those who find it intuitively quite unacceptable may be encouraged to reconsider their attitudes and to ask whether they have the right to exploit the rest of nature without any constraints.

7.   See T. Regan, *The Case for Animal Rights* (University of California Press, Berkeley, 1983).
8.   W. Tucker, "Environment and the Leisure Class," *Harper's*, December 1977, pp. 49–58 and 73–80.

6.3   Ecofeminism

In the 1980s, many feminist thinkers turned their attention to environmental issues, and defined a new term, "ecofeminism."[9] According to one proponent, ecofeminism is "the position that there are important connections – historical, symbolic, theoretical – between the domination of women and domination of nonhuman nature."[10] Its basic premise is that it can construct an environmental ethic by incorporating the problem of patriarchy into our thinking about humans and nature.

Patsy Hallen, an influential American philosopher living in Australia, maintains that domination of women by men parallels our domination of nature.

> Sexism is the expression of a basic pathology of domination and repression. Ecological imbalance is, in part, due to our mistaken belief that we can successfully dominate nature. So sexism (mind and body pollution) is fundamentally linked to ecological destructiveness (environmental pollution).[11]

Hallen believes that because science is our chief means of understanding the environment, we must make science more feminine. She does not believe that women "have a specific essential nature that differentiates them from men" but notes that historically "women's experience . . . has been more involved with nurturing than men's."[12] The experience of women emphasizes wholes and relations, rather than separate individuals and dualisms, and is therefore central to the development of a deeper understanding of environmental ethics. Encouraging more women to become scientists will promote the feminist understanding of the world. The feminist perspective can then weaken the "ideology of detachment and domination" that is central to modern science.

Feminist philosophy emphasizes the ethics of caring, as opposed to the male-oriented ethics of justice. With a concern for caring rather than justice, we can avoid many of the pitfalls of classical ethics. For example, whereas

---

9.   Much of this section is adapted from A. S. Gunn and R. Walker, *Christianity and Environmental Ethics*, a report for the Information and Resource Center, Singapore (1993).

10.   Karen J. Warren, "The Power and Promise of Ecological Feminism," *Environmental Ethics*, vol. 12, no. 2 (1990), p. 125.

11   P. Hallen, "Making Peace with the Environment: Why Ecology Needs Feminism," *Proceedings*, Ecopolitics II Conference, University of Tasmania (1987), p. 111.

12.   Ibid., p. 109.

classical ethicists struggle with the issue of animal suffering, the ecofeminists have no difficulty recognizing that caring for animals is right and good.

Ecofeminism is also a holistic philosophy that concentrates on the place of humans as part of the environment and not as freestanding individuals. Hallen suggests:

> We need to recognize that the whole is more than the sum of the parts
> and that the parts themselves take meaning from the whole. Each part
> is defined by and dependent on the total context. Isolation (as in a lab-
> oratory) distorts the truth because it distorts the whole.[13]

As with transcendentalism and deep ecology, one of the most important aspects of ecofeminism is it spiritual roots. Since the earliest times, people have thought of the earth and nature as feminine. Mother Earth nurtures us. Caring for Mother Earth becomes a simple extension of the responsibility of women.

Ecofeminists have much in common with deep ecologists. For example, they both emphatically reject many assumptions of the Western ethical tradition and invoke a spiritual dimension to environmental ethics.

Although ecofeminists argue that there are connections between the domination of women by men and the domination of non-human nature by people – and that the parallels are so strong that we can begin to develop an environmental ethic based on this association – parallelism is not necessarily proof of applicability to the question. And not everyone will be persuaded by the feminist analysis of environmental exploitation in terms of patriarchy, though certainly there are many parallels.

Ecofeminism suggests an alternative approach to environmental morality – a simple caring for the living and non-living environment that brings us a step closer to the ethic that describes our attitudes toward nature. Whether this caring is gender-specific remains unresolved, however. Perhaps all people, regardless of gender, can learn from the ecofeminist caring approach to the environment.

6.4   The Gaia Hypothesis

The Gaia hypothesis (Gaia was the nurturing earth goddess in Greek mythology) is largely the idea of contemporary biologist James Lovelock,

---

13.   Ibid., p. 123.

who suggested that the earth might be thought of as a single organism (Mother Nature) that lives much like any other organism. This organism struggles to maintain a healthy balance and fights off diseases.[14] Some Gaians interpret the idea literally, taking the view that the earth really is one organism, albeit an unusual one, whereas others see the Gaia hypothesis as a useful metaphor.

If we take this hypothesis literally, humans become simply one part of a whole living organism, the earth, much as brain cells, for example, are a part of the animal. It therefore makes no sense for humans to destroy the rest of the creatures that coinhabit the earth because this would be like destroying our own brain cells. A Gaian would therefore be committed to preserving the earth.

But to many people the application of the Gaia hypothesis to environmental concerns would make our life quite meaningless and ethics would be pointless. If we are not morally responsible individuals with goals of our own and the ability to make our own choices, we may as well just follow our impulses. As in the Calvinist tradition, life is predestined, and we have little to say about what happens to the human race in the long run. This approach to environmental ethics leaves us unsatisfied since it reduces us to unthinking, nonrational organic creatures.[15]

14.   James Lovelock. *Gaia: A New Look at Life on Earth* (Oxford University Press, New York, 1979).

15.   If we accept the notion of Gaia in its literal sense, some interesting extensions can result. We could speculate (and this is pure whimsy) that the earth (Gaia) is still developing and is going through adolescent stages. Most notably, it has not settled on the carbon balance. Millions of years ago, most of the carbon on earth was in the atmosphere as carbon dioxide, and the preponderance of CO_2 promoted the development of plants. But as the plants grew, they soon started to rob the atmosphere of carbon dioxide, replacing it with oxygen. This change in atmospheric gases promoted the growth of animals, which converted the oxygen back to carbon dioxide, once again seeking a balance. Unfortunately, Gaia made a mistake and did not count on the animals dying in such huge numbers, trapping the carbon in deep geologic deposits that eventually became coal, oil, and natural gas.

How to get this carbon out? Invent humans! There is the answer to the age-old question! Our sole purpose on earth is to dig up the carbon deposits as rapidly as possible and to liberate the carbon dioxide. Unfortunately, Gaia again messed up, and did not count on the humans becoming so prolific and destructive – and especially did not count on humans inventing nuclear power, which eliminates the sole purpose for existing. The sheer number of humans (especially if they are not going to dig up the carbon) becomes a problem, much as a pathogenic bacterium causes an infection in an organism. It is not a coincidence that the growth of the human population on earth resembles the growth of cancer cells in a malignancy.

So next Gaia had to limit the number of humans, and she did this by developing "antihumatics" (as in antibiotics), which will kill off a sufficient number of humans and once again bring the earth to a proper balance. The increase in highly resistant strains of bacteria and viruses may be the first indications of a general culling of the human population.

6.5   The Judeo-Christian Tradition

Humans are different from the rest of nature (irrespective of their ultimate reason for existing). We are the only creatures with a developed language and adroit hands with a reversible digit. This difference separates us from the rest of nature, a separation explicit in the Bible, first chapter of Genesis, wherein God tells man to have dominion over all the animals (and presumably all of nature).

> So God created man in his own image, in the image of God he created them; male and female he created them. And God blessed them, and God said to them, "Be fruitful and multiply, and fill the earth and subdue it; and have dominion over the fish of the sea and over the birds of the air, and over every living thing that moves upon the earth."[16]

This oft-quoted passage and its formalization into Christian tradition results in a dualism of people and nature and gives people the right (or expectations) to exploit nature. Whereas ancient religions recognize spirits in nature, spirits that need to be placated, Christianity makes it possible to use natural resources without concern for the spiritual consequences.

In the second chapter of Genesis, however, there is a caveat. God tells people to "have stewardship over all the earth," and places humans in the caretaker role. Humans are still very much apart from nature, however, in that they are the stewards or the gardeners for the earth. Their special responsibility is to take care of the earth for the (absent) owner, providing a possible basis for a useful environmental ethic. It allows humans to prosper while also providing for non-humans and the future. Unfortunately, this has not occurred. In fact, it can be argued that not only is the Judeo-Christian tradition not compatible with an acceptable environmental ethic, it is in fact the *cause* of environmental problems.

Lynn White's 1967 essay "The Historical Roots of Our Ecological Crisis" is undoubtedly the best-known article describing the influence of Judeo-Christianity on the human use of the natural environment.[17] White

---

16. Genesis 1:28.
17. Lynn White, "The Historical Roots of Our Ecological Crisis," *Science*, vol. 155 (March 10, 1967), pp. 1203–1207. Incidentally, White did not originate this view, which, according to Nash (R. F. Nash, *The Rights of Nature* [University of Wisconsin Press, Madison, 1989], p. 51), appears as early as 1894 in an article in *Popular Science Monthly* by Edward Payson Evans. Evans attacked Christianity's anthropocentrism and the "tyrannical mandate" in Genesis, comparing it unfavorably with what he considered to be the biocentric religions of the East. René Dubos (R. Dubos,

(1907–1987), a medieval historian specializing in the history of technology, believed that "Human ecology is deeply conditioned by beliefs about our nature and destiny – that is, by religion," and argued that the modern use of science and technology has its roots in the Judeo-Christian tradition. He did not deny the contributions of Chinese and Arab cultures, but believed that only the Judeo-Christian tradition encouraged the exploitative attitudes that led to "the ecologic crisis." In White's words, "Christianity is the most anthropocentric religion the world has seen," mainly because Christianity inherited from Judaism the belief that God created the world "explicitly for man's benefit and rule: no item in the physical creation had any purpose save to serve man's purposes." Because Islam inherited the same creation story, then on White's interpretation it should be a similarly anthropocentric religion. That they have developed somewhat differently in this respect is partly due to a continued emphasis in Islam on obedience to God as the central and overriding duty, whereas since the Renaissance, Christianity has placed more emphasis on individualism and rights than on duties.

As can be imagined, White's message was not well received by the Christian Church. What White was pointing out was the sorry history of Christianity in encouraging the exploitation of nature and wanton destruction of the environment. As long as we are to have dominion over nature, and as long as God is on our side, who will stop us from treating the rest of nature as something to be conquered and destroyed for human benefit? We note the use of terms such as "animal" and "brute" and "beast" as pejorative, and include in the Anglican *Book of Common Prayer* the phrase "brute beasts which hath no understanding."[18]

Christian doctrine, from the time of Augustine (A.D. 354–430), has placed nature and in particular animals at the pleasure and disposal of humans. According to Augustine:

> Christ himself shows that to refrain from the killing of animals and the destroying of plants is the height of superstition for, judging that there are no common rights between us and the beasts and trees, he sent the devils into a herd of swine and with a curse withered the tree on which he found no fruit.[19]

A Soul Within [Scribner, New York, 1972], p. 157) mentions a 1950 Zen Buddhist source of the thesis. According to Dubos, White only gave the idea "academic glamour."

18.   Andrew Linzey, "For God So Loved the World," *Between the Species*, vol. 6, no. 1 (1990), pp. 12–16.
19.   Quoted in Robin Attfield, *The Ethics of Environmental Concern* (University of Georgia Press, Athens, 1991).

Augustinian Christianity is concerned not with this world but with the next. According to its Christian doctrine, this life is but a preamble to an eternal existence in heaven. Therefore, what we do here matters little as long we make it to that place of eternal bliss. We are free to cut down trees, plow up the soil, kill the animals – as long as we can be assured of salvation.[20] Within Christianity, suggests John Collis, it is possible to love God but to hate his creation.[21]

It is tempting for modernists to believe that science and technology will provide solutions to environmental problems. White believes that this is unlikely, because

> Both our present science and our present technology are so tinctured with orthodox Christian arrogance toward nature that no solution to our ecologic crisis can be expected from them alone.[22]

In White's opinion, we cannot depend on more science and technology. We need to "rethink and refeel our nature and destiny . . . [to] find a new religion or rethink our old one."[23] University of Karachi geographer Iqtidar H. Zaid similarly claims, "Our ecological crisis is in fact a moral crisis and needs a moral solution.[24]

White recognized that Christian thought also contains a more responsible environmental alternative in the writings of St. Francis of Assisi (1181–1226). Francis argued that all of nature is important to God and that to love God is to take care of God's creatures. This is the "stewardship" approach to Christianity, as emphasized in the second chapter of Genesis. White called St. Francis of Assisi "the greatest radical in Christian history since Christ [who] tried to substitute the idea of the equality of all creatures, including man, for the idea of man's limitless rule of creation." Although the Church hierarchy rejected and tried to suppress Francis's views, White celebrates Francis as "a patron saint for ecologists."[25] If such an approach to

20. There are those, of course, for whom "a timeless heaven . . . would be pure hell." Quote from Harold W. Wood, Jr., "Modern Pantheism as an Approach to Environmental Ethics," *Environmental Ethics*, vol. 7, no. 2 (1985), pp. 151–161.
21. John Stewart Collis, *The Triumph of the Tree* (Sloane, New York, 1954).
22. Ibid, pp. 29–30.
23. White, "The Historical Roots of Our Ecological Crisis," pp. 29 and 30.
24. I. H. Zaid, "On the Ethics of Man's Interactions with Nature," *Environmental Ethics*, vol. 3 (1981), p. 36.
25. According to R. Nash, *The Rights of Nature* (University of Wisconsin Press, Madison, 1989), p. 93, this proposal had already been made, by zoologist Martin Bates, in 1960. Nash states that in "1980 the Vatican acted on these suggestions and officially dubbed the Assisi holy man the patron saint of ecologists."

the environment had prevailed instead of Augustine's views, then (according to White) the Western world would have been quite a different place.

Ever since environmentalism became a cause célèbre, the Christian church in the Augustinian tradition has tried to contribute to the search for an environmental ethic. For example, the Church of England created a working group in 1971 "to investigate the relevance of Christian doctrine to the problems of man and his environment." The group's report concluded that

> [creation] exists for God's glory, that is to say, it has a meaning and worth beyond its meaning and worth as seen from the point of view of human utility. It is in this sense that we can say that it has intrinsic value. To imagine that God created the whole universe solely for man's use and pleasure is a mark of folly.[26]

Some Christians extend the Franciscan environmental ethic and assert that humans are not essentially distinct from the rest of nature. For instance, James Nash, the director of the United States Churches' Center for Theology and Public Policy, suggests that the whole of the creation is sacred and that the appropriate human attitude to the earth is love.[27] Nevertheless, these views remain exceptions among major Christian denominations. As British philosopher Robin Attfield writes:

> It is generally agreed [in the Judeo-Christian tradition] that in the Old Testament view nature is not sacred. The Creator and his creation are radically different and it is idolatrous to worship the latter, and so there is nothing sacrilegious in treating creatures as resources for human benefit.[28]

One interpretation of White's paper is therefore that although Christianity is at fault for the sorry state of the environment, it is not necessarily an imperfect religion. Western thought has unfortunately chosen to interpret Christianity in its dualistic (Augustinian) sense as opposed to its holistic (Franciscan) approach.

Of course, we are still left with the question of why the Western world chose the Augustinian approach over the Franciscan. People, after all,

---

26. H. Montefiore, *Man and Nature* (Colliers, London, 1975), p. 67.
27. J. Nash, *Loving Nature: Ecological Integrity and Christian Responsibility* (Abingdon Press, Nashville, TN, 1991).
28. R. Attfield, *The Ethics of Environmental Concern* (Blackwell, Oxford, 1983), p. 25.

choose their own religions. Religious beliefs and dogmas are invented by people, and usually freely accepted by the public if the religion serves some intrinsic spiritual need. If the need is not evident, religions can be imposed, much as Christianity was imposed on the Northern Europeans and the Native Americans. But a religion will not prevail unless it responds to the spiritual needs of the people, and Augustinian Christianity (especially with its blanket permissiveness to pillage nature) was just what the people needed. It therefore makes no sense to blame Christianity as the root cause of our environmental troubles, but neither does Augustinian Christianity seem to be a useful religion around which to build a comprehensive environmental ethic. A revolution would have to occur within Christianity if this source of spirituality can be a foundation for an environmental ethic.

6.6 Asian Religions

There are perhaps thousands of religions that can be labeled Eastern or Asian, and it is not possible to do any of these justice in this discussion. Briefly we mention the Taoist and Confucianist tradition in China, the Zen Buddhist in Japan, and the Hindu religion primarily in India as representative of these religions.

The two oriental religions, Taoism and Confucianism in China, and Zen Buddhism in Japan, have overlapping roots and are here classified generally as Buddhism. The Buddhist tradition considers the whole world in its organic sense, with nothing existing in isolation, and with everything connected to everything else. Humans achieve a harmonious relationship with nature by exhibiting proper humility and caring. Buddhism has as its main tenet the principle of *ahimsa*, "do not destroy life." Buddhism teaches compassion for all of life, including trees, forests, and wildlife. One Buddhist theologian, echoing Jeremy Bentham, puts it this way:

> Buddhism is a religion of love, understanding and compassion, and committed toward the ideal of nonviolence. As such, it also attaches great importance to wild life and the protection of the environment on which every being in this world depends for survival. The simple underlying reason why beings other than humans need to be taken into account is that, like human beings, they too are sensitive to happiness and suffering; they too, just like the human species, primarily seek happiness and shun suffering. The fact that they may be incapable of communicating their feelings is no more an indication of apathy or insensi-

bility to suffering or happiness than in the case of a person whose faculty of speech is impaired.[29]

Such oneness does not hinder development, however. Taoism encourages technological advancement as long as this maintains harmonious balances with nature. What was apparently not evident to the ancient Chinese was that many of the practices leading to the development of higher technology can cause irreparable harm to the soil and animals.[30]

In Japan, the Zen Buddhist movement also stresses the oneness with nature and the merging of self within one's experience. Nature is around us, and in order to master it, we must contemplate and understand it as well as learn to appreciate its beauty. Such appreciation and self-awareness in the Zen tradition leads to fulfillment.

Unfortunately, in both China and Japan the presence of these environmentally enlightened religions does not seem to have prevented massive environmental destruction and disregard for environmental quality. Buddhists apparently contributed significantly to the deforestation in much of Asia to find enough wood to build temples.[31] In China, the communist ideology that seems to have overwhelmed traditional religion is systematically producing massive and irreversible environmental destruction. In Japan, Zen Buddhism has not been enough to give that country a positive environmental record. Following World War II, Japan entered a period of intense industrialization with little concern for the welfare of either its public or the environment, resulting in highly publicized tragedies like the Minemata Bay episode, when hundreds of people were irreparably poisoned by mercury-contaminated fish. Since that time, Japan's environmental awareness seems to have been limited to the protection of the public.

Japan's record in international affairs is equally dismal, including its almost unilateral demand to continue whaling and its continued purchase of lumber from rain forests as well as horns from large African mammals. Ironically, the resurrection of Japan's whaling fleet was encouraged by the United States after World War II, to provide sufficient food for the war-torn nation.

---

29. Lungrg Nomgyal Rimpoche, "The Buddhist Declaration on Nature" (Assisi, Italy, 1986). Quoted in L. G. Regenstein, *Replenish the Earth* (Crossroads, New York, 1991).

30. Joseph Needham, *Science and Civilization in China* (Cambridge University Press, Cambridge, 1954–1984), quoted in Ian Barbour, *Ethics in an Age of Technology* (Harper, San Francisco, 1993).

31. Yi-Fu Tuan, "Our Treatment of the Environment in Ideal and Actuality," *American Scientist*, no. 58 (1970), p. 248.

The sorry record in both China and Japan does not, of course, indict the religious traditions. And certainly both Zen Buddhism and the Taoist tradition may have great possibilities as spiritual bases for an environmental ethic. As before, of course, we are asked to believe some principles on faith alone. That done, however, much good can grow from these traditions. Dr. Nay Htun, a Buddhist professor of environmental engineering, believes that Buddhism has much to offer to the world:

> Compassion and tolerance toward all living beings; respect for all forms of life; harmony with nature rather than the arrogance to conquer it; responsible stewardship of nature for the benefit of present and future generations; less profligate use of resources – these are some of the fundamental attitudes and practices that need to be strengthened. The Lord Buddha recognized these ethical principles and taught and practiced them Himself over two millennia ago. The principles apply even more today.[32]

The other major Asian religious tradition is the Hindu tradition. Like Franciscan Christianity, Hinduism regards humans as stewards for the earth, which was made by God. But unlike Christianity, Hinduism maintains that all animals are incarnations of other living things, including people. Hindus believe that even God was at one time a monkey, a cow, and other creatures, leading to a reverence for certain animals and a prohibition against eating meat.

Hinduism is a religion, a language, and a way of life in India, where more than 80 percent of the population is Hindu. Hindus see humans as a part of the total environment and not as apart from nature. A proverb seems to sum up the Hindu religion: "Do not kill any animal for pleasure, see harmony in nature, and lend a helping hand to all living creatures."[33]

The primary obligation in the Hindu religion seems to be to abide by the rules of the caste into which one is born. Such obedience will assure a higher birth in the next life, whereas disobedience will result in being born below the caste, and truly bad people may be reincarnated as loathsome creatures like snakes and bugs.

The doctrine of *ahimsa*, "do not destroy life," is shared by the Hindu, but not all are vegetarians. Traditionally the cow is sacred to the Hindu, and few would eat beef and still be able to call themselves Hindus.

---

32. Nay Htun, "State of the Environment Today: The Needs for Tomorrow" *Tree of Life*, vol. 19, p. 29. Quoted in Regenstein, *Replenish the Earth.*
33. Regenstein, *Replenish the Earth,* p. 221.

Modern Hindu theologians have promoted the concept of oneness with nature in the Hindu tradition, and promoted care and compassion toward animals. Mahatma Gandhi has been quoted as saying, "If beasts had intelligent speech at their commend, they would state a case against man that would stagger humanity."[34]

While the central tenets of Hinduism require care and compassion for animals and for nature, the actual practice is an imperfect reflection of the faith. Theologian Lewis Regenstein observes:

> The massive commercial exploitation of wildlife in India has had a devastating effect on the local ecology. Tens of thousands of monkeys have been taken from the wild (usually infants captured by shooting the mothers out of a tree) and exported by the animal traffickers for sale to medical researchers, mainly in the United States.
>
> Also very damaging has been the massive commercial killing of snakes, lizards, and other reptiles in India for the Western fashion market, their skins to be made into belts, shoes, handbags and other such goods.[35]

Hinduism also seems to have little effect on how people treat domestic animals. Maneka Gandhi writes:

> Most animals are brought to the city to be killed. They are under the stress of a 200 to 300 kilometer journey in trucks that do not accommodate animals. They break their legs before they arrive. They are left lying there without food and water for a couple of days. Then they kill them [by] slitting their throats and letting them bleed to death. Each animal takes 40 minutes to die.[36]

As with Buddhism, then, although the central theme and tenets of Hinduism promote the care of all creatures and the respect for the environment, the practical everyday experience is quite different from the religious principles.

Even though the Eastern religions seem to be imperfect in their application, these traditions have aroused considerable interest in the West. Some deep ecologists portray the Asian religions as prototypes for formulating a workable environmental ethic. They claim these religions have a deep rever-

---

34. John Paul Fox, *Animals, Cruelty and Kindness* (Salt Lake City, 1979). Quoted in Regenstein, *Replenish the Earth*, p. 225.
35. Regenstein, *Replenish the Earth*, p. 227.
36. As quoted in ibid., p. 232.

ence for life and encourage contemplation and understanding of one's place in the world – motifs not unlike the self-realization and biocentric equality ideals of deep ecology. According to Indian philosopher Ramachandra Guha, however, this is a mistake. Deep ecologists misunderstand Eastern religions, and in their writings,

> complex and internally differentiated religious traditions – Hinduism, Buddhism, and Taoism – are lumped together as holding a view of nature as believed to be quintessentially biocentric.[37]

Guha criticizes the deep ecologists because they ignore the plight of the poor and their need and development, fail to address issues of international trade and exploitation that keep the poor countries from catching up to the rich ones, and, most important, commit the sin of lumping together diverse traditions and misperceive those traditions as biocentric – whereas in reality they are too different to lump together and, indeed, are not biocentric.

6.7   Islam

Although Islam is often grouped with Judaism and Christianity as the "religions of the book," this vast and growing religion deserves individual attention.

In the Islamic tradition, God (Allah) is all-powerful and, as in Augustinian Christianity, has loaned the earth to people, who are to be the caretakers. Islam clearly distinguishes between people and other creatures.

The teachings of the Prophet Mohammed (A.D. 570–632) are recorded in two documents, the Koran (*Qur'ān* in Arabic) and the Hadith. Taken together, they form Islamic law. The Koran is believed to be divine revelations by Allah to Mohammed and are taken as absolute truth by Moslems.

The Koran contains numerous references to animals and the duties of humans to respect nature, but the thrust of these admonitions is uncertain. Most quotations seem to refer to the beauty of the earth and to state that it was placed here for people to enjoy and use.[38] Mohammed did, however, say, "A good deed done to an animal is as meritorious as a good deed done to a human being, while an act of cruelty to an animal is as bad as an act of cruelty to a human being."[39] Echoing the second chapter of Genesis,

---

37.   R. Guha, "Radical American Environmentalism and Wilderness Preservation: A Third World Critique," *Environmental Ethics*, vol. 11 (1989), pp. 71–83.

38.   Regenstein, *Replenish the Earth.*

39.   Ibid.

Mohammed also said. "The world is green and beautiful, and God has appointed you his steward over it."

The Hadith, which is more a collection of Moslem traditions of the Prophet Mohammed than divine revelation, is more definitive about nature. As recorded in the Hadith, Mohammed said, for example: "It behooves you to treat the animals gently. Verily, there are reward for doing good to dumb animals."[40] Mohammed was, according to the Hadith, very sensitive to the sufferings of animals and often taught against their maltreatment. He is quoted as saying that giving food and drink to animals is "among those virtuous gestures which draw us one step nearer to God."[41]

Later Islamic theologians echo the development of Christianity. Abu al-Ala al-Maari (973–1057), the Saint Francis of Assisi of Islam, taught reverence for all animals. He once suggested that it is better to spare a flea than to give charity to the poor.[42]

But while Islamic law contains references to the environment and Mohammed clearly was sensitive to the sufferings of animals, environmental concerns are not central to the religion. University of Karachi geographer Iqtidar H. Zaid states that, from an Islamic perspective, it is

> abundantly clear that God has created the Earth for the service of Man. . . . Man is constantly reminded that the Earth with whatsoever is on its surface, in its interior, and in its atmosphere belongs to Almighty Allah. . . . [M]an is ecologically dominant, but . . . he is not permitted to misuse the bounties of God on Earth and engage in excesses.[43]

An Islamic environmental ethics is therefore possible, depending on the interpretation of "misuse" and "excess." Islam is inherently no different from the Judeo-Christian tradition in its relationship to the environment.

As is the case with Christianity, Buddhism, and probably other organized religions, many Moslems seem to be selective in the parts of the Koran they choose to obey. Regenstein observes:

> Anyone who has ever visited or lived in a Moslem country knows that animals in those areas of the world are hardly treated with the compassion and respect required by Islamic law. Indeed, observers are often

40. Ibid.
41. Ibid.
42. Ibid.
43. I. H. Zaid, "On the Ethics of Man's Interactions with Nature," p. 43.

shocked by the scenes of animal suffering so commonly seen in the
Moslem world – overloaded donkeys being whipped and beaten, stray
dogs and cats being starved and kicked, sheep being sacrificed in the
streets. Little regard is shown for the environment, and even critically
endangered species of wildlife are avidly hunted.[44]

Nevertheless, Islam is probably more conductive to an environmental
ethics than many other of the world religions. It is interesting that in
Malaysia, where a moderate form of Islam is the official religion, attempts
are being made to promote concern for future generations and nature as
inherent to Islam.

6.8   Animistic Religions

Most ancient religions, including Polynesian and Native American religions,
are animistic and recognize the existence of spirits within nature. In such
religions the spirits do not take human form, as in the Greek, Roman, or
Judaic religions. They simply are within the tree, or the brook, or the sky. It
is possible to commune with these spirits, talk to them, feel close to them,
and thereby feel close to nature. As Lynn White recounts,

> every tree, every spring, every stream, every hill had its own *genius loci*,
> its guardian spirit . . . [thus] before one cut a tree, mined a mountain,
> or dammed a brook, it was important to placate the spirit in charge of
> that particular situation, and to keep it placated.[45]

Christian theologians usually emphasize that, like it or not, modern
humans are no longer a part of nature. In Carl Jung's words:

> As scientific understanding has grown, so our world has become dehu-
> manized. Man feels himself isolated in the cosmos, because he is no
> longer involved in nature and has lost emotional "unconscious iden-
> tity" with natural phenomena. These have slowly lost their symbolic
> implications. Thunder is no longer the voice of an angry god, nor is
> lightning his avenging missile. No river contains a spirit, no tree is a life
> principle, no snake the embodiment of wisdom, no mountain cave the
> home of a great demon. No voices now speak to man from stones,

---

44.   Regenstein, *Replenish the Earth.*
45.   L. White, "The Historical Roots of Our Ecological Crisis," p. 120.

plants, and animals, nor does he speak to them believing they can hear. His contact with nature has gone, and with it has gone the profound emotional energy that his symbolic connection supplied.[46]

In many animistic religions the killing of an animal, such as a deer or a bear, requires the proper appeasement of that spirit. Cutting down a tree requires an explanation to that tree (spirit) as to the reason for cutting it down, and the explanation better be a good one or the spirit will haunt the cutter long after the event. In some societies, such as the Sami (formerly known as Laplanders), large stone outcroppings in the treeless tundra become sacred places where spirits live, and where it is possible to converse with them.

Some modern environmentalists romanticize the ancients and look to animism with admiration, as if these people had some wisdom that we no longer possess. Unfortunately, this is probably unwarranted adoration. White notes that "ever since man became a numerous species he has affected his environment notably."[47] For example, primitive fire-drive hunting methods exterminated Pleistocene megafauna and perhaps created the world's great grasslands. And according to a recent study, Polynesians wrought the complete destruction of Easter Island before European contact.[48]

The near extermination of the plains buffalo in the mid-1800s has always been decried as a Western crime. And yet it appears that the Europeans actually hired Native Americans to do much of the butchering. George Catlin was a writer and painter of the Native Americans in the 1830s, and describes his first-person experience this way:

> [I was told] that only a few days before I arrived [at Fort Pierre, in present-day South Dakota] (when an immense herd of buffalos had showed themselves on the opposite side of the river, almost blackening the plains for a great distance), a party of five or six hundred Sioux Indians on horseback, forded the river about mid-day, and spending a few hours amongst them, recrossed the river at sun-down and came into the Fort with *fourteen hundred fresh buffalo tongues* [emphasis in original] which were thrown down in a mass, and for which they

---

46.   C. S. Jung, *Man and His Symbols* (Doubleday, New York, 1964), quoted in J. Hinchcliff, "Toward an Environmental Ethic; or, Back to the Old, Old Story," *Proceedings,* Toward an Environmental Ethic (Colloquium, University of Waikato, New Zealand, 1977), p. 39.

47.   L. White, "The Historical Roots of Our Ecological Crisis," p. 18.

48.   J. Flenley and P. Bahm, *Easter Island, Earth Island* (Thames & Hudson, London, 1992).

required but a few gallons of whiskey, which was soon demolished, indulging them in a little, and harmless carouse.

The profligate waste of the lives of these noble and useful animals, when, from all I could learn, not a skin or a pound of the meat (except the tongues), was brought in, fully supports me in the seemingly extravagant predictions that I have made as to their extinction, which I am certain is near at hand.[49]

The idea that non-Western or precapitalist societies are necessarily conservationist is more myth than historical reality. Nevertheless, their approach holds considerable promise in the formulation of our own environmental ethic. Consider the possibility of believing that all things (animals, plants, inanimate objects) have spirits. How would our actions toward the non-human world change if we respected these spirits? We would not, for example, cut down a tree or even pull out a weed without first (perhaps tacitly) apologizing to the plant and explaining the necessity of our actions. In effect, such an approach would create an environmental ethic of respect for everything (humans included).

6.9   Modern Pantheism

Modern pantheism is an updated version of both transcendentalism and animism. For modern pantheists, "the ecosystem can be viewed as a scientific revelation of God."[50] Because God is no longer some abstract entity but can be discovered in daily observations, the study of nature becomes a religious undertaking. To modern pantheists, the objective of the religion is not to seek eternal life in some heaven through the study of scriptures such as the Bible, but to seek God on earth through the study of nature.

According to Harold Wood, modern pantheism offers three approaches to achieving the oneness with God – the way of knowledge, the way of devotion, and the way of works.

---

49. George Catlin, *North American Indians: Being Letters and Notes on their Manners, Customs, and Conditions, Written during Eight Years' Travel amongst the Wildest Tribes in North America, 1832–1839*, vol. 1 (George Catlin, London, 1880), p. 288. Quoted in Roderick Frazier Nash, *American Environmentalism: Readings in Conservation History* (McGraw-Hill, New York, 1990), p. 31.
50. Dennis G. Kuby, "When I Stand at an Open Grave," *Ecology and Religion Newsletter*, no. 37 (January 1977), quoted in Harold W. Wood, Jr., "Modern Pantheism as an Approach to Environmental Ethics," pp. 151–161.

Reflecting Emerson's transcendentalism, the way of knowledge in modern pantheism is the study of nature. As Wood describes it:

> The astonishment with which we gaze upon the starry heavens, or the microscopic life in a drop of water; the awe in which we trace the workings of energy throughout the world; and the reverence with which we search for truths which we know we may never fully grasp, is the essence of a religious feeling.[51]

Modern pantheism contrasts with most religious worship in that devotion involves not the praise of a deity but simply the observation of nature.[52] Again Wood:

> A sense of miraculous is hardly difficult to achieve when contemplating the wonders of the universe. As such, a devotion centering on the primeval life stream, both externally and internally, is an authentic form of prayer or meditation.[53]

Vern Crawford describes how such devotion can take place:

> Sitting quietly in a forest, feel the earth beneath you. Be alert to it. Touch the soil, and look down at your sensitive hand. Then say to yourself, and to the whole universe: "Who is here? It is *I!* and I am *here!*"[54]

Pantheist devotion, as with animists, can also celebrate parts of nature, such as the hydrologic cycle. Even though we do not ascribe the various functions of the hydrologic cycle to various gods as the ancient animists did, the process is no less wonderful and awesome. If people considered such central aspects of nature with proper reverence, they would not wantonly destroy natural systems and processes. As Wood points out, modern pantheists would view the acid rain controversy on a totally different plane. They would see acid rain as an attack on the wondrous hydrologic cycle, and oppose it. Thus, pantheists move from reflection, through devotion, to action.

The modern pantheist views the earth in a religious way and practices

---

51. Wood, "Modern Pantheism," pp. 151–161.
52. Or as F. W. Nietzsche puts it: "I cannot believe in a God who wants to be praised all the time." Quoted in Jon Winokur, ed., *The Portable Curmudgeon* (New American Library, New York, 1987), p. 120.
53. Wood, "Modern Pantheism," pp. 151–161.
54. Vern Crawford, "Communication with Nature: How Is It Done?" *Days Afield Perspective*, vol. 1 (1983), quoted in ibid.

conservation and reservation not out of some self-motivated interest but out of a sense of religious duty toward the environment. In the "way of works," the pantheist tries to act in the best interest of the ecosystem, including the humans who inhabit it. Pantheists do not need to explain these actions, any more than other religions depend on logical explanations for their rituals. The ritual of modern pantheists is to work toward a better environment and to love one another.

Modern pantheism seems to speak to our basic needs, but again requires a religious experience and a foundation. In this case, however, one must see a divine hand in the workings of nature. Then knowledge, devotion, and good works follow from this basic belief.

Is this the source of the environmental ethic we seek? John Stewart Collis has an optimistic view of our future. He writes:

> Both polytheism and monotheism have done their work. The images are broken, the idols are all overthrown. This is now regarded as a very irreligious age. But perhaps it only means that the mind is moving from one state to another. The next stage is not a belief in many gods. It is not a belief in one god. It is not a belief at all – not a conception in the intellect. It is an extension of consciousness so that we may *feel* God.[55]

6.10   Conclusions

If we are to accept any of the approaches to environmental ethics based on spirituality, we must accept a priori some tenet or principle without rational proof. In deep ecology, we need to accept the idea of self-realization and biocentric equality; in Christianity we need to accept the divinity of Jesus, in the animistic religions we need to accept the existence of spirits in nature, and so on. Once we are able to do so, we can build the environmental ethic on that foundation.

Of course, the same may be said of any other ethic. For example, a belief in liberal–democratic forms of society requires that one accept universal human rights, a tenet that cannot be demonstrated to be true. Thus, it may not be too much to ask of us to believe in some form of spirituality as the foundation of an environmental ethic.

---

55.   Collis, *The Triumph of the Tree.*

This may seem like an unsatisfactory conclusion, however. Although we may be environmentally concerned, we want to justify this concern by an appeal to something beyond an article of faith or our personal preferences. Like Carole, who would not run over the box turtle, we want to be able to rationalize our actions.

But since none of the other approaches to environmental ethics (codes of ethics, classical ethics, and extensionist ethics) seem to offer knock-down arguments for a basis for an environmental ethic, and if we reject the accounts of environmental ethics based on spirituality, what are our options?

Basically, we have three options:

1. Stop worrying about environmental ethics. One way to avoid spending lots of time and effort on a problem is to decide that it isn't really a problem after all. Sometimes this can be very healthy and liberating: the realization that we are never going to be astronauts, great cooks, or truly enlightened persons can free us to concentrate on realistic and achievable goals. So, given the complexity and scale of global environmental problems, is there any point in worrying about our impact on the earth? The earth is not going to fall apart during our own or our children's lifetimes. After that, the people of the future can take care of themselves.

As with any material possession, we would not want to be deprived of clean air or uncontaminated water. As for the non-human world, the dualism of people/nature can prevail, and in the absence of reasons to the contrary, we do not care about the welfare of non-human creatures or the quality of the environment other than what would be of immediate use to us. This is fine, so long as we can adopt this position sincerely without stifling our feelings and our critical faculties.

The second alternative is:

2. Abandon any attempt to develop an environmental ethic and just adopt some ideas based on our inner feelings.

This is also fine, so long as we are prepared to accept that environmental ethics is no more than a sincere personal commitment to the environment. Otherwise, it can be a very frustrating position. For example, suppose we are concerned about the unspeakable suffering of calves when they are grown for white veal and we are convinced that this is a cruel and despicable practice. But if we have no rational argument to convince those who eat veal to share our view, we have no basis to persuade them to stop supporting the veal industry.

There is a third alternative:

3. Adopt an environmental ethic that makes the most sense to us based on our understanding of the world and on our evaluation of the ethical arguments advanced in this book. We would have to admit that there are no "drop dead" arguments for environmental ethics, and thus we could never conclusively prove to anyone that our values are ones that everyone should adopt. But we would also acknowledge that we can make some reasonable assumptions and from these can flow an acceptable environmental ethic. Reason and evidence can take us a long way, facts can alert us to problems, and logic can test whether our ideas and assumptions work. Reason enables us to derive conclusions, to show how ideas are connected, and to achieve consistency.

But as the preceding chapters of this book demonstrate, the accumulation and analysis of factual information can help but not solve the problem of deciding environmental issues. Once some facts are known, the effect of more facts may not change anyone's mind, since the argument is no longer about the facts but about different values. But where do these values come from?

Some values no doubt are the result of our educational experience, especially parental, school, and media. But there must be more to it than that. One is tempted to think that these values are spiritual or even universal, but as with other matters of faith, no amount of rational argument will convince anyone otherwise.

When Aldo Leopold was a young man working for the Forest Service, one of his first assignments was to help in the eradication of the wolf in New Mexico. One day he and his crew spotted a wolf crossing a creek and instantly opened fire, mortally wounding the animal. Leopold scrambled down to where it lay and arrived in time to "watch the fierce green fire dying in her eyes."[56] This experience was significant in Leopold's recognition of the value of predators in an ecosystem and his eventual acceptance of the intrinsic value of all nature.

Did Leopold's spirituality see the fire in the wolf's eyes, or was that spirituality transferred from the wolf to Leopold? How much does nature itself, in all of its myriad forms, create a sense of unity and belonging to the greater environment? Thinking about this is difficult and confusing, but perhaps this is what makes us so special, and what it means to be human.

56. A. Leopold, *A Sand County Almanac* (Ballantine, New York, 1966; originally published in 1949), p. 138.

Suggested Supplemental Readings for Chapter 6

1. *Ecological Feminism,* Karen Warren
2. *The Judeo-Christian Stewardship Attitude Toward Nature,* Patrick Dobel
3. *Sanctity of Life in Hinduism,* O. P. Dwivedi
4. *Islamic Environmental Ethics,* Mawil Y. Izzi Deen (Samarrai)
5. *Respect for Nature,* Gary Snyder.

7

Incorporating Environmental Ethics into Engineering

Engineering decisions affect not only people but also the non-human environment. That engineers should be concerned with their effect on the environment is a very new idea for the "people-serving profession." Nobody has (yet) declared engineering the "nature-serving profession." Engineers work for other people and have a special responsibility toward their employers and clients. But who, as in Dr. Seuss's *Lorax,* speaks for the trees?[1] Who is the Lorax who tries to warn us of stupid and unthinking acts that so often devastate our environment?

One approach for engineers is to withdraw into their professional shells and allow the "environmentalists" to keep an eye on the world. Most industries are scared to death of the much-storied "little old lady in tennis shoes." And engineers might be tempted to abrogate their responsibility for environmental decisions by suggesting that unless someone is vehemently opposed to their actions, the actions must be environmentally sound. But engineers are often the only people with the knowledge of potential environmental harm and the professional authority to command attention – if anything, this gives them *more* environmental responsibility.

Some engineers use the similarly dubious technique of the "public hearing" to place a stamp of approval on their plans. This method reached perfection with the U.S. Army Corps of Engineers. First, the engineers revealed their plans to construct some large facility. Then the people who would benefit from this facility, such as real estate developers and local merchants, would comment at a public hearing. The hearings were often attended only by those people who had the most to gain from the expenditure of public money. Once initiated and publicly approved, the projects gained momentum and attained a life of their own, quite apart from any rational grounds for their continuation.

---

1.   T. Geisel (as Dr. Seuss), *The Lorax* (Random House, New York, 1971).

One example of many such engineering projects is the "strange ditch in the Everglades."[2] In concert with real estate developers and farmers, the Corps constructed a large ditch to drain a freshwater marshland near Lake Okeechobee in the Florida Everglades. None of the planning documents noted that this drainage would allow severe saltwater intrusion and would wipe out a large percentage of many rare and endangered birds, such as the wood stork, the great white heron, and the bald eagle. This tragedy could have been avoided. All disinterested engineering and hydrologic skill cautioned against the drainage ditch, but the Corps proceeded anyway, inflating the benefits by increasing the land values as needed to cover the escalating costs of construction. The project had no apparent worth, and caused considerable harm, but its tremendous bureaucratic momentum guaranteed its completion. But if an engineer, using any sincere environmental ethic, had spoken early and strongly against the project, it might never have gained such momentum. That engineer could have served people and nature at the same time.

Because engineering works affect the public, much of what the engineer can and cannot do is specified by law. When a whole society is committed to an environmental ethic, it will codify these values into law. In the United States, the most important environmental legislation is the National Environmental Policy Act of 1969. In New Zealand, the most comprehensive piece of environmental legislation is the Resource Management Act of 1991. Each of these acts has affected the practice of environmental engineering in the respective countries, and each defines the role of professional engineers in environmental protection. We discuss them here to illustrate the role of engineers in environmental decision-making.

7.1   The National Environmental Policy Act

In part because of the inability of engineers to assimilate the effect of their projects on the environment into their design procedure, the U.S. Congress passed and President Richard Nixon signed into law the National Environmental Policy Act (NEPA). This legislation, in tandem with the Endangered Species Act passed four years later, provided legal ammunition for environmental groups to battle government projects. The chief weapon was a little-discussed section of NEPA, Section 102, that mandated that any government project which significantly affects the environment must have an

2.   George X. Sand, *The Everglades Today: Endangered Wilderness* (Four Winds Press, New York, 1971).

Environmental Impact Statement (EIS). The purpose of this statement was to force the engineers, planners and managers at least to consider the effect their plans would have on other species, on the visual environment, and on long-term conservation and use of resources.

At first engineers dismissed and even ignored the EIS. They figured they would write some generic disclaimer and attach it to every report, as with today's job nondiscrimination notices. But they were foiled by the courts. For example, the New Hope Dam Project in North Carolina (later the B. Everett Jordan Dam), which eventually cost well over $100 million, was first graced with a mere seventeen-page EIS. It did not take the court long to determine that this was totally inadequate. The Corps was forced to reconsider the question and develop a comprehensive statement outlining the impact of this project on the environment. Although the B. Everett Jordan Dam eventually was constructed, the protracted fight over the environmental effects undoubtedly caused other Corps of Engineers projects to be reevaluated and perhaps even scrapped.

Over the years, the form of the EIS has changed, and the procedures have been streamlined, but the positive effect remains. It is no longer possible, as it was in the Everglades project, simply to ignore the bald eagles, for example. Back then, engineers did not even consider the survival of birds an engineering problem. Now, with the EIS, engineers must include the welfare of other species in the planning process.

This should not suggest that the EIS has successfully introduced an environmental ethic into the engineering profession. Although the EIS helps a great deal by forcing engineers to expand their thinking on environmental problems, it is not a total solution. There are several reasons for this:

1. The EIS applies only to some federal projects. Most large projects by the Department of Defense, for example, like weapons testing, do not require EIS's.
2. The EIS applies only to projects that significantly affect the environment. Each government agency can determine what is and what is not significant.
3. Although most states have some form of EIS, few states enforce them effectively. Thus, most state-funded engineering projects lack this safeguard.
4. No EIS is required for privately funded projects. Thus, if a large agricultural firm wants to strip the surface of a huge peat deposit in order to carry on high-intensity farming (with almost certain contamination of the groundwater), there is no mechanism (other than construction per-

mits and other regulations) that will force the engineers and entrepreneurs to consider the effects on the environment.

5.   There is no procedure presently available for making the value judgments once the factual information is compiled and analyzed.

Consider this last point for a moment. Suppose the Corps finds that flooding a marshland would most probably cause the destruction of habitat for some bird, say, the bald eagle. The EIS lists the pros and cons of the project, including alternatives to the project, and at the end a decision is to be made. "Yes," the project is to go forward, or "no," it is to be stopped. Who decides that, and how? It is easy to calculate the *positive* effect of a project by just estimating the increase in property values or some other benefits. But the negative value, that of environmental destruction, is very difficult to quantify in economic terms. Some approaches have been proposed for using economics for calculating environmental costs, such as asking what one would be willing to pay *not* to do something like destroy a wetland, but these techniques are inadequate and anthropocentric. And yet, the engineers want the EIS to be quantitative so that a clear decision can be made.

In practice, most engineers arrange their calculations of environmental effect in such a way that the outcome reflects exactly what they wanted to do in the first place. Since there are no standard methods for writing the statements, engineers can easily manipulate the judgmental data to obtain the desired result.

The most important criticism of the EIS, however is this:

6.   Even though the EIS might show severe damage to the environment, and even though everyone (including the engineers) agrees that this is the case, there is no guarantee that the project will be scrapped or modified.

This point is important, so it deserves repeating. The EIS, even though it might be highly negative, can not automatically stop or even change a project. The decision to do so is reserved for the political process.

If this is true, why should engineers worry about values and ethics? Are not engineers, then, simply hired hands, doing the bidding of the society that employs them?

The response to that question is that engineers *do* decide policy in many ways. On a local and mundane level, engineers make seemingly minor decisions every day that affect all of us. For example, the routing of a local highway across a stream that may be a rare habitat for the endangered Venus's-flytrap is an engineering decision that will get no press whatever, much less receive White House consideration. And on a national level, the mere initiation of a major project will result in a momentum difficult to reverse. Thus, it is not enough simply to defer engineering decisions to public comment or environmental oversight. Engineers should introduce environmental concerns into project planning before the plans become public documents.

7.2   The Resource Management Act

The New Zealand Resource Management Act of 1991 is, to our knowledge, the only legislation with explicit references to environmental ethics, including direct reference to intrinsic and aesthetic values and natural character. This act is a comprehensive piece of environmental management legislation, and emphasizes the traditional values and rights of the Maori, the indigenous people of New Zealand, especially under the Treaty of Waitangi. Under this treaty, which was signed in 1840 between an agent of the British Crown and a number of Maori chiefs, the latter agreed to yield the powers of central government to the Crown in return for legal recognition of Maori rights to control their own affairs and their fisheries, land, and other resources. Although it must be admitted that the Treaty was not respected by successive New Zealand governments, it has been explicitly recognized in recent years by statute law, the courts, and governmental agencies.

An important feature of the act is that courts and the agencies that are hearing resource management applications for permission to commence or continue activities that have environmental impact – in particular coastal activities, such as effluent discharge, beach use, offshore work, and coastal subdivisions – must give priority to the areas addressed in Part II of the act, which includes intrinsic values of ecosystems, the natural character of the environment, Maori cultural and traditional values, indigenous flora and fauna, public access, aesthetics, and amenity value, as well as social and economic factors. If these values sound familiar, they should; the government officials who drafted the legislation have admitted that they were very much influenced by various readings on environmental ethics, in particular the writings of Aldo Leopold.

Although this legislation is too recent for us to gauge its ultimate effect on the New Zealand environment, a recent decision suggests that it may make major changes. The circumstances and arguments leading to this decision demonstrate the effectiveness of this act.

A small coastal town applied for a continuation of its permit to discharge

sewage to an ocean outfall. The effluent was to be treated to secondary standards, and apart from the risk of a pipeline rupture there was no evidence of
risk to human health or environmental quality as measured by such standards as BOD, indicator microorganisms, and heavy metals. But representa-
tives of the local Maori people objected very strongly to the discharge
because, according to their values, human waste must never be discharged
into water without first passing through the earth. A pipeline that passes
over shellfish beds and fishing areas therefore pollutes those waters, as well
as the discharge point and adjacent areas. Treatment levels are irrelevant:
treated sewage is still sewage. Thus, the Maori are unable to use a large
coastal and marine area for food gathering, fishing, and recreation. Obviously, they are suffering an economic loss, but the evidence presented to the
hearing emphasized the shame they felt in being unable to offer the traditional hospitality of seafood to visitors because of the spiritual pollution of
the food.

The Maori people were not the only objectors to the resource permit, but
they were the only ones to insist that the town convert to an entirely land-
based sewage disposal system. The government panel, acting according to
the Resource Management Act, had all the input it needed to make its decision. On the basis of the Maori representation the panel granted the town
only a short-term permit and ordered the town to develop, in consultation
with the Maori people, a disposal scheme that would in practice mandate a
land-based system. Is environmental ethics – here in the form of respect for
Maori values – indeed and at last making a difference?

7.3   Conclusion: The Ethical Engineer

We begin this book by stating that ethics, as we conceive it, does not provide
neat formulae for the solution of ethical problems and the development of
ethically justifiable systems for decision making. If ethics is conceived of as
a very detailed and specific list of rules, the engineer who is looking for guidance in a particular situation might expect to look up the relevant rule and
apply it. But we believe this way of doing ethics is often unhelpful. For one
thing, it will work (if it works at all) only if there already *are* settled and
detailed rules, on which everyone agrees. Environmental ethics is still at an
early stage of development; it would be stupid and presumptuous for the
authors or anyone else to dictate a complete list of environmental ethical
principles for engineers.

Even if there were such a list of rules, it could never be complete for all time, because novel unforeseen situations will arise, and there will always be a need to select the appropriate rule, to interpret it, to apply it to the situation, and to deal with potential conflicts between rules.

What we try to provide in this book is an account of the main resources that are available within the broad Western cultural tradition and the world's main religious traditions. We emphasize that there is more than one way of doing environmental ethics. No way is perfect, but that is acceptable because nothing is perfect, even in engineering design. Any ethical position requires that assumptions be made, and any ethical position can be critically evaluated from different perspectives. An ethical engineer is not one who subscribes to *the* ethical position, however, because in our view there is no one perfect ethical theory. Rather, there are a number of ethical perspectives, which work out successfully in the sense that they contribute to the flourishing of humans and the rest of nature, and whose prescriptions and values have a great deal in common.

This is just as well, because in many ways there is globally an increasing diversity of values. For instance, the number of different religions in English-speaking countries has multiplied over the last few decades. Traditional secular humanism and its extensions have become widely accepted, and new forms of thought have arisen, including feminism and deep ecology. Concurrently, there has been an increased awareness and acceptance of cultural diversity. Without the ability to understand different cultural and ethical perspectives, our collective search for a healthy environment cannot succeed.

In this book we demonstrate that it is possible to develop at least some major elements of environmental ethics within each perspective. Christians need not set aside their Christianity when they consider environmental issues. Feminists likewise can apply their feminist values to environmental concerns. In other words, we can justify our environmental attitudes in many ways, whether we believe that we are God's stewards, or that living things and nature have rights, or that ecosystems are intrinsically valuable, or that we should care for and nurture all things; or that we must protect natural systems for the sake of future generations.

This consensus should not surprise us. Perhaps the analogy to slavery can help again. In social ethics, many different perspectives *agree* that slavery is wrong, whether one argues that slavery violates natural rights, or that slavery violates the ethics of caring, or that it is economically inefficient, or that it is not optimal on utilitarian grounds, or that it is incompatible with an ethic of respect for persons, and so on.

Engineers are in a unique position to make a difference in the care and nurturing of our planet. Their understanding and appreciation of environmental ethics is thus of great importance. Private and governmental clients, if they are to get things done, must use engineering know-how. They do not necessarily have to use other experts, such as sociologists, philosophers, epidemiologists, or even planners, but they must hire engineers. Thus, engineers have a responsibility to their professional roles in the broadest sense – to introduce alternatives, concepts, and values that the client may otherwise never consider. Truly professional engineers infuse ethics into their decision making. With a growing population, accelerated technological development, and an increasing pressure on the natural environment, environmental ethics will become an ever greater part of the engineer's role in society. The ethical engineer *can* make a difference.

Part II

Supplemental Readings for Chapter 1

The Kepone Tragedy

WILLIAM GOLDFARB

The tragedy of human greed vs. the environment is perhaps best illustrated by the events surrounding the production of Kepone in Hopewell, Virginia, by the Allied Chemical Corporation and its jobbers. In this statement of the facts surrounding the case, it becomes clear that both Allied and its contractors considered only the advantages of personal enrichment and did not care what the effect on the environment (or in this case, even on humans who worked for them) would be.

William Goldfarb is an environmental lawyer with the Department of Environmental Science at Cook College, Rutgers University, New Brunswick. This excerpt is from a paper presented at the Institute of Government, Eastern Kentucky University.

Hopewell (population approximately 24,000) is an industrial city located on the banks of the James River in southern Virginia. Calling itself "the chemical capital of the South," Hopewell has actively solicited the large chemical concerns. Consequently, Firestone, Hercules, Continental Can, Allied Chemical have located chemical plants in Hopewell.

In 1949, the initial batch of 500 pounds of a pesticide named Kepone was produced by Allied, and two patents for the process were awarded to it in 1952. Allied did not consider Kepone to be a major pesticide. In fact, Kepone production never exceeded 0.1 percent of America's total pesticide production, and its sales were less than $200,000 annually over a sixteen year period. Kepone was intended primarily for export to Europe for use against the Colorado Potato Beetle, and to South America to control the Banana Root Borer.

Excerpted from William Goldfarb, "Kepone: A Case Study," *Environmental Law*, vol. 8 (1978), p. 645. Used with permission.

Before moving to commercial production, Allied inaugurated an extensive series of toxicity tests involving Kepone. Such tests were necessary in order to obtain registration under the Federal pesticide laws, which Allied actually received in 1957. The tests were conducted by various consultants, including the Entomology Department of the College of Agriculture at Rutgers University (now Cook College). The results of this research revealed Kepone to be highly toxic to all species tested: it caused cancer, liver damage, reproductive system failure, and inhibition of growth and muscular coordination in fish, mammals, and birds. Upon being presented with the test results, Allied voluntarily withdrew its petition to the Food and Drug Administration for the setting of Kepone residue tolerances for agricultural products.

Kepone is a chlorinated hydrocarbon pesticide, a chemical relative of DDT, Aldrin/Dieldrin, and Mirex (all of which have been banned by the United States Environmental Protection Agency – EPA). As such, Kepone is a contact poison, capable of being absorbed through the skin or cuticle; it is lipophilic (fat soluble), but insoluble in water; it is persistent in the environment; and it will bioaccumulate in the fatty tissues of the body. The exact mechanism by which chlorinated hydrocarbons kill target pests is uncertain. What is known is that they are nerve poisons, interfering with the transmission of electrical impulses along nerve channels. The results of contact with Kepone are loss of control over muscular coordination, convulsions ("DDT-like tremors"), and eventually death.

Despite the unfavorable toxicity test results, Allied deemed Kepone ready for commercial production, and contracted with the Nease Chemical Company of State College, Pennsylvania, to produce it for Allied. The relationship between Allied and Nease lasted from 1958 through 1960. Allied entered into a similar arrangement with Hooker Chemical during the early '60's.

By 1966 even more negative test results had been associated with Kepone, but Allied nevertheless decided to manufacture Kepone on an increased basis in its own Semi-Works facility in Hopewell. In preparation for production an area supervisor of the Semi-Works was asked to develop a production manual. The production manual was to contain operating and safety instructions for the production process. The supervisor naturally consulted available toxicity research results, and his recommended precautions reflect the test findings. At Allied, Kepone spills and dust were closely controlled and workers wore safety glasses and rubber boots and gloves. Allied's Kepone operations were directed by William Moore until 1968, and thereafter by Virgil Hundtofte.

Prior to preparation of the production manual, there had been no recorded case of human exposure to Kepone to the level of acute poisoning. Allied apparently discounted such a possibility, regardless of the documented adverse effects of Kepone on animals. However, a witness for the United States at the trial testified that Allied should have suspected "that the same symptomology would be induced in man if exposed to Kepone."

In 1970, the Federal government resurrected the Refuse Act Permit Program, which required all industrial discharges into navigable waters to obtain permits from the U.S. Army Corps of Engineers. The Allied complex at Hopewell had three discharge pipes directly into the stream called Gravelly Run, a tributary of the James River. One of these pipes originated at the Semi-Works where Kepone was manufactured. The Refuse Act Permit (RAP) application was discussed by Allied's plant managers and their assistants, who found themselves on the horns of a dilemma. Allied was discharging Kepone process wastes without treatment of any kind, and the installation of pollution control equipment would be expensive. Moreover, planning was being conducted for the construction of a regional sewage treatment plant which would treat the wastes of all industries in Hopewell, by the municipal treatment plant would not be complete until 1975 at the earliest. What should Allied do during the construction period?

Allied decided to list the Semi-Works discharge as a temporary phenomenon which would be discontinued within two years. In such cases, Federal regulations allowed for a short form RAP application requiring only that the discharge be identified as a "temporary discharge." (Allied gratuitously added that it was unmetered and unsampled.) Thus, neither Kepone nor two plastics products (TAIC and THEIC) also manufactured at the Semi-Works were listed by Allied on its RAP application, even though Allied quite clearly did not intend to terminate production at Hopewell.

In 1972 the RAP expired, but a new permit program had been enacted – the National Pollutant Discharge Elimination System (NPDES) permit program under the Federal Water Pollution Control Act Amendments of 1972 (FWPCA). The NPDES permit program is administered by EPA instead of the Corps of Engineers.

EPA requested data on the nature, volume, and strength of Allied's discharges, and again the dilemma manifested itself. One of Hundtofte's assistants prepared an option memorandum, outlining three strategies which Allied might follow: 1) to do nothing and hope for a lack of enforcement by EPA; 2) to divert the Semi-Works effluent to another outfall pipe for which a permit had been obtained; or 3) to slowly improve the Semi-Works effluent so as to "buy time" until completion of the municipal system. None of these options, however, entailed the disclosure by Allied of its Semi-Works effluent. The last of these options was selected, and again Allied submitted data to the Federal government describing the Semi-Works discharges as unmetered, unsampled and temporary outfalls. As a result, between 1966 and 1974 Allied discharged untreated Kepone and plastics wastes into Gravelly Run without revealing the nature of its discharges to the Federal government.

In 1973 Allied underwent a corporate reorganization, during which control of the Semi-Works facility was transferred from the Agricultural Division to the Plastics Division. The former was superseded in expectation of its impending move to new facilities in Baton Rouge, Louisiana. Virgil Hundtofte, Plant Manager of the Agricultural Division at Hopewell, and William Moore, Research Director, made plans to retire from the company rather than to relocate. (Hundtofte had been with Allied in Hopewell since 1965, Moore since 1948).

One effect of the reorganization was a reorientation of production priorities among the products manufactured at the Semi-Works. Kepone production had decreased steadily, but THEIC, which had been manufactured in small quantities for eighteen years, suddenly found a lucrative market calling for a doubling of production. THEIC and Kepone shared certain production equipment, and with the surge in demand for THEIC a decision was made in 1973 to "toll" Kepone production. Tolling is a common arrangement in the chemical industry whereby another company performs processing work for a fee or a "toll" and then returns the final product. The keynote of a tolling arrangement is that during the processing period legal title to the materials and product remains with the supplier, in this case Allied.

In January, 1973, when the decision to toll Kepone was divulged, William Moore saw his opportunity to remain in Virginia and continue in the Kepone manufacturing business. He immediately contacted Hundtofte, who had recently resigned from Allied and gone to work for a fuel oil distributor. Moore and Hundtofte agreed to form a corporation and bid for the Kepone tolling contract. On November 9, 1973, Life Science Products Company (LSP) was incorporated under the laws of the Commonwealth of Virginia. Moore and Hundtofte were the only shareholders, directors and officers of LSP. Less than a month later, the tolling agreement between Allied and LSP was signed. Allied had solicited bids from Hooker Chemical, Nease Chemical, Velsicol, and LSP, but LSP's bid was by far the lowest:

54 cents per pound for 500,000 pounds of Kepone. Nease Chemical (which, it may be recalled, manufactured Kepone for Allied from 1958 through 1960) declined to bid, but responded that if it chose to bid on the contract it would cost Nease 30 cents per pound for waste disposal alone. Hooker (Nease's successor) bid $3.00.

The details of the tolling agreement are important because the question of Allied's responsibility for LSP's illegal acts loomed large at the trial. The contract provided that Allied would supply – at its own expense – all of the raw materials for Kepone production, with the title to remain in Allied. Within certain broad limits, Allied would determine the monthly production rate of Kepone, which would be packed in Allied containers and transported in Allied trucks. Allied also agreed to pay LSP's taxes, other than corporate income taxes. LSP was to receive between 32 and 38 cents per pound for 650,000 pounds or more of Kepone. Through a capital surcharge arrangement, Allied was to pay for all of LSP's approved capital expenditures, whether for production or pollution control, except for land and building. If LSP closed for pollution violations during the first year of the contract, Allied had the option to purchase LSP's assets for $25,000. And if the contract was terminated by either party for any reason, LSP agreed to refrain from producing Kepone for anyone else.

The relationship between Allied and LSP was only partially defined by the tolling agreement. Moore and Hundtofte promised Allied that they would not dispose of their shares in LSP without Allied's consent. Moreover, Allied assisted LSP in many ways: to obtain loans and equipment from outside sources; to meet temporary cash deficits; to augment fuel supplies during the oil embargo; to attain greater efficiency by the use of Allied facilities. Most importantly, LSP's effluent was sampled and analyzed by Allied personnel after Virginia ordered such testing. Before that (up to October, 1974), LSP had tested its effluent by a visual check – if the effluent was cloudy, the presence of suspended Kepone was indicated.

Allied officials regularly toured the LSP plant, and were informed by mail of the waste disposal problems which LSP faced almost from its inception. Whereas Allied had discharged the residues of its Kepone production process directly into tributaries of the James River, LSP decided to discharge into Hopewell sewer system, despite the fact that the regional treatment plant was still under construction. By this means, LSP could avoid having to apply for an NPDES permit, which is not required of "indirect discharges" (discharges into municipal treatment systems). LSP contacted

C. L. Jones, Director of Hopewell's Department of Public Works, for permission. At that time, Hopewell possessed a primary waste treatment plant – a series of settling tanks without biological or chemical treatment other than disinfection and sludge digestion. Such a rudimentary system could not degrade Kepone, but would merely divide Kepone influent between outfall pipe and sludge. Jones, who had been Plant Manager of Allied's Hopewell Semi-Works prior to Hundtofte, recommended to Hopewell's City Manager that Hopewell accept LSP's wastes. (In order to assure that no damage would be done to the sludge digester at the treatment plant, LSP was asked by Hopewell to meet a pretreatment standard of three parts per million of Kepone.) Thus, LSP became the only industry in Hopewell allowed to discharge into the municipal sewerage system.

LSP commenced operation in March, 1974, and almost immediately, large quantities of Kepone began flowing into Hopewell's treatment plant. In October, a State inspector discovered that the sludge digester at the plant was inoperative, and his investigation revealed LSP to be the source of contamination. Prior to the plant breakdown, the State was apparently unaware that Kepone was being discharged into the Hopewell system because Hopewell's application for an NPDES permit for its treatment plant (filed on October 10, 1973 – about a month before plug-in permission was granted to LSP) made no mention of any industrial discharge to the municipal system. Moreover, additional information filed by Hopewell in July, 1974, did not include notice of LSP's discharges. When the State brought the situation to the attention of LSP and Hopewell officials, LSP's discharges were not prohibited even though the pretreatment standard was being violated, but a study was commenced to determine a "safe" effluent limit for Kepone. Finally, in June of 1975, a more restrictive pretreatment standard was imposed upon LSP (0.5 parts per billion). EPA, which had been informed of the situation, agreed to this compromise, even though it had earlier recommended a stricter effluent limitation. In order to meet this standard LSP had to further pretreat its wastes and hold its discharges in tanks until such time as the discharge would not violate the pretreatment standards (i.e. even out the flow). Allied had participated in the negotiations among LSP, Hopewell, Virginia, and EPA; and Allied opted to pay for the necessary pollution control equipment. Allied and LSP then began to discuss the capital costs of expanding Kepone production to 2,500,000 pounds per day in order to meet an increasing demand in the European market. (From the inception of LSP, Allied had constantly requested increased Kepone production.)

However, even after the new equipment was installed, the pretreatment standard was violated in nineteen out of twenty-one samplings.

On July 7, 1975 as LSP was preparing for increased Kepone production, one of its employees visited an Hopewell internist named Chou, complaining of tremors, weight loss, quickened pulse rate, unusual eye movements, and a tender, enlarged liver. Such symptoms were not unusual among LSP employees, but were generally dismissed as "the shakes" – a necessary price to be paid for the $5.00 per hour wage they received. Although about twenty physicians had been consulted during the sixteen months of LSP's existence, only Dr. Chou suspected a connection between the ailments and the workplace environment. After questioning his patient and taking a blood sample, Dr. Chou forwarded the sample to the Center[s] for Disease Control in Atlanta, where he knew that an analysis for Kepone could be performed. The tests disclosed that the blood sample contained 7.5 parts per million of Kepone, an astounding concentration to be found in human blood. Federal doctors then contacted the Virginia State epidemiologist, Dr. Robert Jackson, who quickly arranged a meeting with Hundtofte and Moore. Similar meetings had previously been requested by the Virginia State Department of Labor and Industry, but LSP had been successful in postponing them. Even though a former LSP worker had filed a complaint with the United States Department of Labor, that agency had also failed to inspect the LSP factory. Only one Federal official, an EPA pesticide inspector, visited LSP before July, 1975; but he was not authorized to enter the production area. Other media under EPA jurisdiction – air and water pollution – had been delegated to State and local officials. At various times representatives of the Virginia Air Pollution Control Board, Water Pollution Control Board, and State Health Department had visited LSP; but they were not responsible for inspecting LSP's working conditions.

When Jackson toured the plant he was appalled: "Kepone was everywhere"; conditions were "incredible"; workers were virtually "swimming in the stuff." Workers wore no protective equipment, nor had warning signs been posted. Seven out of ten production workers present had "the shakes" so severely that they required immediate hospitalization. On July 25, 1975, LSP voluntarily closed out its operation under threat of a closure order by the Virginia Department of Health. Further investigation divulged seventy-five cases of acute Kepone poisoning among LSP workers and high levels of Kepone in the blood of some of their family members. Moreover, there was found to be massive contamination of air, soil, and especially water in the vicinity of the plant. As a result, the State of Virginia closed one hundred miles of the James River and its tributaries to fishing.

(*Editors' note:* Allied was eventually convicted on more than a thousand counts of pollution, and paid a substantial fine. Nobody went to jail and nobody was fired. In the next annual report following the payment of the fine, Allied reported that this wasn't a big thing and they could easily cover the fine from operating revenue.)

The Hooker Memos

CBS NEWS

This is a direct transcript from the CBS program *60 Minutes*. Hooker Chemical Company, a subsidiary of Occidental Petroleum, appeared to have been disposing of toxic materials into the ground for years and polluting the groundwater. Such contamination is considerably more serious than the pollution of surface waters (as was the case in Hopewell) since the contamination of groundwater is for all intents and purposes permanent. What is interesting in this article is the apparent ignorance of the company executives as to what was really taking place. One almost feels sympathy for Mr. Baeder, the president of Hooker. The fault seems not to be so much his as in the inability of middle managers to act on serious but clearly defined problems that may not have been in the company operations manual. Management was not impressed by appeals for reason, even when written in terms of potential financial liability to the company. For middle managers to heed the concerns expressed in the memos from Mr. Edson would have required a special appreciation of and concern for the environment – an enlightened environmental ethic.

MIKE WALLACE: Tonight, "The Hooker Memos" provide a rare inside look into an American company that some have called a "corporate outlaw." The company is the giant Hooker Chemical. Even before these memos surfaced, Hooker's image was in trouble. Back in 1977, residents of Love Canal in Niagara Falls had to evacuate their homes because soil contaminated by Hooker's hazardous waste had seeped into their basements. Last October, Hooker said it would spend $15 million to clean up damage caused by its chemical pollution of a lake in the State of

Transcript of "The Hooker Memos," as broadcast by "60 Minutes," 16 December 1979. © 1979, CBS News. Used with permission.

Michigan. And now the State of California is investigating Hooker on charges that the company's Occidental Chemical Plant at Lathrop, California, for years dumped toxic pesticide waste in violation of state law, polluting the ground water nearby. Hooker officials deny the charges, but Hooker memos seem to say the company knew what it was doing and, nonetheless, just kept on doing it.

The man who wrote these memos, Robert Edson, is Hooker's environmental engineer at their Lathrop plant. He wouldn't speak to us, but his memos speak for themselves.

On April the 29th, 1975, Robert Edson wrote: "Our laboratory records indicate that we are slowly contaminating all wells in our area, and two of our own wells are contaminated to the point of being toxic to animals or humans. THIS IS A TIME BOMB THAT WE MUST DE-FUSE."

June 25th, 1976, a year later, there was this: "To date, we have been charging waste water . . . containing about five tons of pesticide per year to the ground . . . I believe that we have fooled around long enough and already overpressed our luck."

A year after that, on April 5th, 1977, Edson was still writing memos. "The attached well data" he says, "shows that we have destroyed the usability of several wells in our area. If anyone should complain, we could be the party named in an action by the Water Quality Board. . . . Do we correct the situation before we have a problem or do we hold off until action is taken against us?" And on September 19th, 1978, more than a year later, Edson again pointed out to his management: "We are continuously contaminating the ground water around our plant."

And this is that plant, the huge Occidental Chemical complex where Bob Edson's memos say chemical waste was being dumped and was poisoning the ground water. Back in 1968, the Water Quality Control Board for the region came to an understanding with Oxy Chem about what this plant would be permitted to discharge as waste. Oxy Chem agreed to discharge no substance that would cause harm to human, plant or animal life. But back then, the water board thought that Oxy Chem was dealing only with fertilizers and nontoxic fertilizer waste. It turned out that was not the case. What the water board did not know, had no idea of, was that another kind of waste was being dumped in here, toxic waste from pesticides, which were also manufactured over at that plant but were nowhere mentioned in Occidental's agreement with the water board. Hooker Chemical, itself a subsidiary of the giant

Occidental Petroleum, headquarters in Houston, Texas, and we wondered if Hooker's president Don Baeder, would talk to us about the Lathrop Plant and those Edson memos. He would. But he maintained that the water board in California had known about and had okayed Occidental Chemical's discharge of pesticide wastes. And he told us that Robert Edson simply didn't know that.

DON BAEDER: Mike, it's unfortunate, but Mr. Edson didn't take on the environmental assignment until 1972, so he was unaware of this visit by the state to our facilities. And . . .

WALLACE: May I see that a second?

BAEDER: Yes, you may. I wish you'd read that last paragraph, too.

WALLACE: "We wish to thank Occidental Chemical for the spirit of concern and cooperation in which this problem was met and to commend you for the thoroughness of your approach." Signed Charles Carnahan, the executive officer of the California Regional Water Quality Control Board. And this is dated 4 September 1970.

BAEDER: Does that sound like a company that was not concerned with the environmental problems back in 1970, Mike?

WALLACE: I'm at a loss, then, to understand why your own Mr. Edson would say some five years later, "Recently published California State Water Quality Control laws state we cannot percolate chemicals to ground water. The laws are extremely stringent about pesticides. And to date, the Water Quality Control people don't know about our pesticide waste percolation." This is your own man.

BAEDER: Yes, and Mr. Edson . . .

WALLACE: Chief environmental engineer for Oxy Chem at Lathrop.

BAEDER: Mr. . . . and Mr. Edson is a good man. And Mr. Edson somehow felt it necessary to . . . to steel the management to action with these kind of statements. But they didn't need steeling.

WALLACE: June 25th, 1976, a memo from R. Edson to A. Osborne. Who's A. Osborne?

BAEDER: I believe he's another engineer at the plant.

WALLACE: "For years," he says, "we've dumped waste water containing pesticides and other ag–chem products to a pond southwest of our plant. Our closest neighbor's drinking water well is located less than 500 feet from the subject percolation pond and, fortunately for the management, no pesticide has yet been detected in his water. I personally would not drink from his well." This is Edson, this fellow that you respect so.

BAEDER: I do.

WALLACE: This fellow who should have known what . . . Why didn't you show him this letter that you've shown me?

BAEDER: I don't . . . I think the important thing, Mike, is . . .

WALLACE: But . . .

BAEDER: . . . what actions that we took to . . .

WALLACE: No, wait just a second. Why didn't you, someplace along the line, say, "Hey, Bob. Don't you know that the California Water Quality Control Board back in 1970 sent us this letter? They – you're . . . you're . . . you're talking through your hat, Bob Edson, because the California people knew all about this all the time."

BAEDER: The fact that certain of our people in our organization didn't know about it isn't . . . isn't germane. The fact is the Water Quality Control Board did know about it.

WALLACE: But Charles Carnahan, the man who signed that letter Don Baeder has shown us, Charles Carnahan, the executive officer of the water board, told us that when he wrote that letter, he had no idea that Occidental was dumping toxic pesticide waste.

CHARLES CARNAHAN: The waste that they were dumping we figured was the waste from their fertilizer manufacturing.

HARRY MOSES: How do you feel about it now that you've learned differently?

CARNAHAN: Well. I feel kind of stupid. I feel stupid because I think that we took them at . . . on good faith and we were fooled.

WALLACE: But Don Baeder, who still maintained that the water board did know, also insisted that no harm was done, and that, he said, is what is really important.

BAEDER: No one has been hurt at Lathrop. No one. From our discharges, no one has been hurt and no one will be. No one will be.

DR. ROBERT HARRIS: When you contaminate ground water, you do so, for all practical purposes, irreversibly.

WALLACE: Dr. Robert Harris is an expert on toxic chemicals, recently appointed to the President's Council on Environmental Quality.

DR. HARRIS: The reason is that ground water moves very slowly, and that these contaminants that are now present in the ground water will continue to be there, will continue to migrate towards population centers, toward individual wells, for decades to come. This water will not be cleansed easily.

WALLACE: This past summer, the folks here at Oxy Chem drilled 12 test wells to find out just how contaminated the ground water had become. Five of the wells showed quantities of a pesticide called DBCP, and this well, number seven well, contained it in excessive amounts. Enough, according to Dr. Harris, to increase the risk of dying from cancer by as much as 25 percent if the water from this well were consumed over a normal lifetime.

DR. HARRIS: These are not small concentrations of these particular chemicals.

WALLACE: Is it possible they just didn't know, that the state of the art at the time that this kind of thing went on was such that they didn't know what they were doing?

DR. HARRIS: They should have known. They should have known that DBCP had and did cause sterility in laboratory animals more than a decade ago, and they should have known that DBCP was being tested at the National Cancer Institute, and that preliminary results showed that it was a very potent carcinogen as early as 1973. This is information that was available to the general public, and I assume was . . . was made available to Hooker.

BAEDER: As soon as we found out it caused sterility, as soon as we found out it caused cancer, those operations were shut down.

WALLACE: When was that?

BAEDER: That was in 1977, I believe, the summer of '77.

WALLACE: Shortly after you took over as president?

BAEDER: Yes.

WALLACE: All right. And you don't want to go back into the business of making DBCP?

BAEDER: We are not going back into the business of making DBCP.

WALLACE: You don't want to go back in?

BAEDER: I don't want to.

WALLACE: Well, then, how come I have a memo here, dated December 11th, 1978, to D. A. Guthrie from J. Wilkenfeld, subject: "Re-entry to DBCP Market"? Why would you . . . why would your company inquire into the possibility of – quote – "re-entry into the DBCP market" if you don't want it? You're the president. This happens in 1978.

BAEDER: Mike . . .

WALLACE: Why . . . why would they be exploring it, Mr. Baeder?

BAEDER: Mike, if the government permits it and we develop safe . . . safe systems for handling it to eliminate any exposure, any potential harm to people, it . . . it could be and will be produced. It's an important . . . it's an important chemical to agriculture of California. Now just because if something is mishandled it is toxic or it causes cancer does not mean that that same material cannot be handled safely. And I can

assure you that the State of California would not allow anyone to produce that under risk of these kinds of . . . of . . . of problems without
markedly changing the procedures under which it was produced.

WALLACE: It is my understanding, according to this memo here, that the
State of California has rejected requests for permission to use the mate-
rial, and the government of Mexico has shut down the two manufac-
turing plants for DBCP that were operating in that country. And yet,
your Acting Vice President for Environmental Health and Safety, Mr.
David Guthrie, says "We've reviewed the proposal for re-entry into the
DBCP business and we have no environmental health objections. Jerry
Wilkenfeld has no technical objections." Who's Wilkenfeld?

BAEDER: Mr. Wilkenfeld is a . . . is responsible for environmental health
safety at the Occidental level.

WALLACE: For environmental health and safety?

BAEDER: At the Occidental . . .

WALLACE: Let me read you what he says.

BAEDER: You did read it.

WALLACE: "Assume" . . . Oh, no, I haven't read the whole thing.

BAEDER: Oh.

WALLACE: "Assume that 50 percent of the normal rate for these people
exposed may file claims of effects from the exposure. Determine the
number of potential claims for sterility and cancer, based on the insurance department's or legal department's estimate of the probable average judgment or settlement which would result from such a claim.
Calculate the potential liability, including 50 percent for legal fees and
other concise . . . contingencies." One gets the impression that profits are more important to the Hooker Chemical Company than care for
human health.

BAEDER: We went out of DBCP as soon as we found out it presented any
harm or exposures to the people or the workers.

WALLACE: And a year ago you were back in the business of looking at the
possibility of going back into the business.

BAEDER: People are looking at it, but we are not into it. And that would have
been a very deliberate corporate decision to go back into it. And I . . .
look, Mike, before I'm a president, I'm a human being.

WALLACE: Uh-hmm.

BAEDER: We would not have gone back into it. Again, you're – you're dealing
in studies that people make. We make a lot of studies. Why do you ham-
mer us on something that might have happened but hasn't happened?

WALLACE: The only reason I'm hammering it is: Is this the way that America does business?

BAEDER: America looks at many options in doing business. It looks at many options.

WALLACE: And one of the options is, can we afford . . .

BAEDER: No. No. Mike, there is a risk in making almost an . . . there's a risk in making drugs. Mike, your drug people look at this same thing in every pharmaceutical drug they put on the market. There is a risk. There . . .

WALLACE: Do you know how this . . . this memo ends up? It says, "Should this product" – DBCP – "still show an adequate profit, meeting corporate investment criteria, then the product should be considered further." That's the bottom line.

BAEDER: Mike, I tried to tell you that profit is not the primary consideration. Mike –

WALLACE: It's your own memo.

BAEDER: Mike, it's a memo by a . . . a young man in the corporation. It's not the policy of the corporation. Young people do not set policy. Mike, I tried to tell you that over the last three years in environmental health and safety, we've spent more than a hundred and thirty million dollars. That's more than our profits have been.

WALLACE: I wonder why.

BAEDER: Why? Because . . .

WALLACE: Yeah. Why have you had to?

BAEDER: . . . because environment is important. Mike, we're a . . . you know, we . . . we have a concern for our people.

WALLACE: Fine . . .

BAEDER: . . . we have a concern for our neighbors, and that's why we're spending this money.

REPRESENTATIVE ALBERT GORE (D-Tennessee): Well, I want to believe him, and I hope that what he says is true. I hope that this company and the chemical industry generally will accept the very great public responsibility that it has.

WALLACE: Congressman Albert Gore is a member of the House subcommittee that has been holding hearings on hazardous waste disposal, and he is genuinely alarmed by the extent of the problem.

REPRESENTATIVE GORE: Two or threes decades from now, in many parts of the country we'll be facing widespread shortages. This is a very precious resource which must be conserved and protected, yet we are systematically poisoning it by dumping 80 billion pounds every year into the ground. It was formerly thought that the ground was just like a big sponge that could soak up all the poison that we could pour into it, but it's not the case.

WALLACE: How bad is Hooker Chemical, Congressman? Is it one of the worst or just in the middle of the catalogue of chemical corporate outlaws?

REPRESENTATIVE GORE: The . . . the problem is industrywide. They are every day making hard calculations, just like this company did, as to whether or not the risk to other Americans is worth it for them to make a good deal more money.

BAEDER: The State of California, most states, permit certain discharges to the environment, because it's their belief that they can be tolerated. When we learn otherwise, we change. The problem we have is that, with today's knowledge, there's no question we would have done something differently. The real issue is: Should we be judged by today's knowledge on past practices? And I think that's the issue. I think the American people have got to understand that.

GEORGE DEUKMEJIAN: The evidence appears to us to indicate very clearly that they have done this willfully and knowingly.

WALLACE: George Deukmejian, California's attorney general, is currently asking that Hooker Chemical pay millions of dollars in fines. But are fines an effective deterrent? Or, we wondered, would jail terms for corporate officers whose companies break the law be a more effective deterrent?

DEUKMEJIAN: I don't really feel that we need to have additional laws as much as we need to enforce the laws that we do have. And I think that if we do that, I feel that we will be able to control this type of practice. Also, when you get into the criminal law area, you're talking about the need, I suppose, to prove a criminal intent, and that might be difficult, if not im . . . impossible, to do in a case of this type.

WALLACE: What do you think about the suggestion of criminal penalties for corporate outlaws?

BAEDER: Mike, I think that criminal penalties for criminal acts are completely justified.

WALLACE: Is it a criminal act to poison, sterilize, whatever, if it can be proved that it was done with knowledge?

BAEDER: Mike, if it can be proved that causing sterilization by violating the law, by . . . by operating in a way that you know is unsafe . . .

WALLACE: Hm-mmm?

BAEDER: . . . I would say that's a criminal act.

WALLACE: And who should serve the time or pay the fine?

BAEDER: I think the people that are involved?

WALLACE: The top man or . . .

BAEDER: I think the buck always stops at the top, sir.

WALLACE: Since we filmed this report, there has been this late development. California Attorney General George Deukmejian will file a lawsuit this coming week against Hooker Chemical. The suit will ask Hooker to pay fines and cleanup costs in excess of $15 million for the damage caused by their Lathrop plant.

(*Editors' Note:* The Occidental Petroleum Company is presently in the process of spending millions of dollars to extract contaminated water from the ground so that the plume of pollution will not extend any farther.)

The Bunker Hill Lead Smelter

CASSANDRA TATE

In the case of Hooker Chemical management, conscious decisions were made to ignore the warnings of impending environmental disaster. It was not a case of not knowing. It was a case of deliberately choosing to act in what we would now consider an environmentally destructive manner. But what if there were true conflicts where several values of equal importance conflicted? What if the problems were not of the nature of environment vs. profits?

In the Bunker Hill lead smelter case, just such a dilemma is faced by the management of a primary materials industry. What is not stated in the article is the nature of the business and how precariously such industries remain operational in the United States. Not only is there a question of job safety, but a question of employment opportunity in the community. And in some cases, even the survival of the community. Dilemmas such as this are daily fare for engineers who have attained management positions.

Lead is one of the oldest, most ubiquitous, most toxic, and most thoroughly studied substances used by man. Lead can damage the nervous system, the kidneys, and the reproductive organs. At sufficiently high levels, it causes convulsions, coma, and death. The effects of lead begin at low levels, so that although there may be no observable symptoms, damage may have been done, and it may be irreversible.

Excerpted from C. Tate, "The American Dilemma of Jobs, Health in an Idaho Town," *Smithsonian*, vol. 12, no. 6 (September 1981). Used with permission.

"There's nothing wrong with my kids," says Kathy Kriedeman flatly. Kriedeman, her husband, Lowell, and their two children live in a small, tidy house on the main street of Smelterville, a community with a post office, two taverns, about 840 citizens, and some of the highest concentrations of lead in the Kellogg [Idaho] area. Her children, now aged 9 and 13, both had lead levels higher than 70 micrograms when tested during the Centers for Disease Control (CDC) survey. Kriedeman – a large, sturdy woman in her early thirties, with mild blue eyes – has refused to have them participate in any of the numerous follow-up surveys since then. She has declined several offers to have them tested for neurological or psychological abnormalities. "I don't like all these people poking at my kids, sticking their noses in where they don't belong."

Kathy stayed home until her children were out of the toddler stage. Then she joined her husband at "the Bunker," getting a job in the smelter.

The smelter had not hired women – except for a brief period during World War II – until the early 1970's when it was opened up in response to pressure from the Equal Employment Opportunity Commission. Kriedeman was one of the first to be hired. The Occupational Safety and Health Administration (OSHA), the federal agency that sets and enforces standards for workplace health and safety, has measured lead levels up to 7.37 milligrams – 7,370 micrograms – per cubic meter of air over an eight-hour period.

Respirators were available to smelter employees at the time. but their use was not strictly enforced. Kriedeman rarely wore hers. "They're too hard to breathe through, especially when you're working." Ten months after beginning her job, she discovered she was pregnant and asked for a transfer. "There was all that smoke coming directly up in my face. I was getting sick from it." But, she says, her request was denied, so she quit her job. A week later, she miscarried. She says her doctor, since deceased, told her that the miscarriage had probably occurred because she had been working in the smelter.

Kriedeman returned to work several months later. She worked until April, 1975, when Bunker Hill announced that women capable of bearing children would no longer be employed in he smelter because of "medical information which indicated lead might cause fetal complications." This meant that a woman of child-bearing capability would have to undergo sterilization in order to be eligible to work in the smelter. The 29 women then at work were transferred to other jobs. Eighteen of them, including Kriedeman, promptly filed charges of sex discrimination, to which an exasperated James H. Halley, Bunker Hill's president at the time, responded:

"So, am I supposed to put women in the plant, or am I supposed to keep them out? Which is the most moral thing to do? If we don't put women in the smelter that's going to mean fewer jobs for women. If we put women in the smelter, and she gets pregnant, we're liable to have a mentally retarded person born who otherwise would have been normal."

Supplemental Readings for Chapter 2

The Existential Pleasures of Engineering

SAMUEL FLORMAN

The antitechnology movement in the 1960s was aided in great part by the inability of engineers to articulate what they did for a living and why. To many, the ills of our society were *caused* by the engineers, who were responsible for such items as nuclear bombs, pesticides, and airplane crashes. Samuel Florman, an eminent civil engineer, rose to the defense of engineering in a widely read book *The Existential Pleasures of Engineering.* He points out first that engineering is *fun.* "What a startling word." He continues, "Engineering is *fun,* and similar to the creative arts in providing fulfillment." In the section of his book reproduced below, he argues for this aspect of engineering, and then goes on to show that the nature of engineering has been misconceived. "Analysis, rationality, materialism, and practical creativity do not preclude emotional fulfillment; they are pathways to such fulfillment. Engineering is superficial only to those who view it superficially. At the heart of engineering lies existential joy."

Perhaps many engineers will agree that engineering is indeed fun. But what is this *existential* business? Isn't this totally alien to the spirit of engineering? Not so, insists Florman, as long as one interprets existentialism as the search for inner truth, the elimination of external influence, and the disenchantment with conventional creed and delusions. Engineers, in their search for the best solution to technical problems, are, according to Florman, free of such externalities, and thus have existential freedom to perform their tasks.

This all sounds attractive, until the question of social and environmental responsibility is raised. Florman's philosophy, in our opinion, does not deal adequately with such questions. As perhaps the most gruesome example to illustrate this problem, consider the fact that the gas ovens in Nazi extermination camps were designed by *engineers*. They were simply given a job to do, and they went about doing it, free of any social concern. The construction of unneeded, uneconomical, and environmentally disastrous dams, waterways, and other facilities by the Corps of Engineers differs both in scale and in substance, but not in philosophy. Engineers have fun – yes – but often at the expense of other humans or the environment.

The first and most obvious existential gratification felt by the engineer stems from his desire to change the world he sees before him. This impulse is not contingent upon the need of mankind for any such changes. Doubtless the impulse was born from the need, but it has taken on a life of its own. Man the creator is by his very nature not satisfied to accept the world as it is. He is driven to change it, to make of it something different. Paul Valery, in his poetic drama, *Eupalinos*, has expressed this impulse with a romantic flourish:

> The Constructor . . . finds before him as his chaos and as primitive matter, precisely that world-order which the Demiurge wrung from the disorder of the beginning. Nature is formed and the elements are separated; but something enjoins him to consider this work as unfinished, and as requiring to be rehandled and set in motion again for the more special satisfaction of man. He takes as the starting point of his act the very point where the god left off . . . the masses of marble should not remain lifeless within the earth constituting a solid night, nor the cedars and cypresses rest content to come to their end by flame or by rot, when they can be changed into fragrant beams and dazzling furniture.

This desire to change the world is brought to a fever pitch by the inertness of the world as it appears to us, by the very *resistance* of inanimate things, to use the concept expressed by Gaston Bachelelard in *La Terre et les Reveries de la Volonte:*

> The resistant world takes us out of static being . . . And the mysteries of energy begin . . . The hammer or the trowel in hand. We are no longer alone, we have an adversary, we have something to do . . . All these *resistant* objects . . . give us a pretext for mastery and for our energy.

The existential impulse to change the world stirs deep within the engineer. But it is a vague impulse that requires particular projects for its expression. Here the engineer cannot help but be enthralled by the countless possibilities for actions that the world presents to him. In *A Family of Engineers,* Robert Louis Stevenson has told of the allure that the profession of engineering had for his grandfather:

> . . . the perpetual need for fresh ingredients stimulated his ingenuity . . . The seas into which his labors carried the new engineer were still scarce charted, the coast still dark. . . . The joy of my grandfather for his career was as the love of a woman.

The engineer today, for all his knowledge and accomplishment, can still look out on seas scarce charted and on coasts still dark. Each new achievement discloses new problems and new possibilities. The allure of these endless vistas bewitches the engineer of every era.

For many engineers, the poetic image of seas and coasts can be taken literally. Water and earth are the substances that engaged the energies of the first engineer – the civil engineer. Civil engineering is the main trunk from which all branches of the profession have sprung. Even in this age of electronics and cybernetics, approximately 16 percent of American engineers are civil engineers. If we add mining, basic metals, and land and sea transportation, fully a quarter of our engineers are engaged in the ancient task of grappling with water and earth. Civil engineering has traditionally included the design and construction of buildings, dams, bridges, railroads, canals, highways, tunnels – in short, all engineered structures – and also the disciplines of hydraulics and sanitation: water supply, flood control, sewage disposal, and so forth. The word "civil" was first used around 1750 by the British engineer, John Smeaton, who wished to distinguish his works (most notably the Eddystone Lighthouse) from those with military purposes. The civil engineer, with his hands literally in the soil, is existentially wedded to the earth, more so than any other man except perhaps the farmer. The civil engineer hero of James A. Michener's novel, *Caravans,* cries out, "I want to stir the earth, fundamentally . . . in the bowels." The hydraulic engineer hero of Dutch novelist A. Den Doolaard's book, *Roll Back the Sea,* stares across the flood water rushing through broken dikes and feels "a strange and bitter joy. This was living water again, which had to be fought against."

Living Water.   Nature, which appears at one moment to be inert and resistant, something which the engineer is impelled to modify and embellish, in the next instant springs alive as a flood, a landslide, a fire, or an earthquake, becomes a force with which the engineers must reckon. Beyond emergencies and disasters, through the environmental crisis of recent years, nature had demonstrated that she is indeed a living organism not to be tampered with unthinkingly. Nature's apparent passivity, like the repose of a languid mistress, obscures a mysterious and provocative energy. The engineer's new knowledge of nature's complexities is at once humbling and alluring.

Another dichotomy with which nature confronts the engineer relates to size. When man considers his place in the natural world, his first reaction is one of awe. He is so small, while the mountains, valleys and oceans are so immense. He is intimidated. But at the same instant he is inspired. The grand scale of the world invites him to conceive colossal works. In pursuing such works, he has often shown a lack of aesthetic sensibility. He has been vain, building useless pyramids, and foolish, building dams that do more harm than good. But the existential impulse to create enormous structures remains, even after he has been chastened. Skyscrapers, bridges, dams, aqueducts, tunnels – these mammoth undertakings appeal to a human passion that appears to be inextinguishable. Jean–Jaques Rousseau, the quintessential lover of nature undefined, found himself under the spell of this passion when he came upon an enormous Roman aqueduct:

> . . . walked along the three stages of this superb construction, with a respect that made me almost shrink from treading on it. The echo of my footsteps under the immense arches made me think I could hear the strong voices of the men who had built it. I felt lost like an insect in the immensity of the work. I felt, along with the sense of my own littleness, something nevertheless which seemed to elevate my soul; I said to myself with a sigh: "Oh! that I had been born a Roman!" . . . I remained several hours in this rapture of contemplation. I came away from it in a kind of dream . . .

The rapture of Rousseau for the "immensity of the work" survives in the midst of our most bitter disappointment with technology. A 1964 photo exhibition at The Museum of Modern Art in New York, entitled "Twentieth-Century Engineering," brought home this truth to a multitude of viewers. The introduction to the exhibition catalogue directed attention to the fact that the impact of enormous engineering works is sometimes enhanced by the "elegance, lightness, and the apparent ease with which difficulties are overcome," and sometimes by the opposite, the monumental extravagance that appears when "the engineer may glory in the sheer effort his work involves." Ada Louise Huxtable of the *New York Times* reacted to the show with an enthusiasm that even the proudest of civil engineers would hesitate to express:

> It is clear that in the whole range of our complex culture, with its self-conscious aesthetic kicks and esoteric pursuit of meanings, nothing comes off with quite the validity, reality, and necessity of the structural arts. Other art forms seem pretty piddling next to dams that challenge mountains, roads that leap chasms, and domes that span miles. The kicks here are for real. These structures stand in positive, creative contrast to the willful negativism and transient novelty that have made so much painting and literature, for example, a kind of diminishing, naughty game. The evidence is incontrovertible: building is the great art of our time.

"The kicks here are for real." And if they are for real to the observer of photographs, imagine what they are like to the men who participate in creating the works themselves. *Roll Back the Sea*, the Dutch novel already mentioned, has a scene describing the building of the Zuyder Zee wall which gives some slight taste of the excitement surrounding a massive engineering work:

> The great floating cranes, dropping tons of still clay into the splashing water with each swing of the arm. Dozens of tugboats with their white bow waves. Creaking bucket dredges; unwieldy barges; blowers spouting the white mass of sand through a long pipeline out behind the dark clay dam; and the hundreds of polder workmen in their high, muddy boots. An atmosphere of drawing boards and tide tables, of megaphones and jingling telephones; of pitching lights in the darkness, of sweat and steam and rust and water, of the slick clay and the wind. A dike in the making, the greatest dike that the world had ever seen built straight through sea water.

The mighty works of the civil engineer sometimes appear to be conquests over a nature that would repel mankind if it could. Thus Waldo Frank perceived the Panama Canal slashing through the tropical jungle:

> Its gray sobriety is apart from the luxuriance of nature. Its Willfulness is victor over a voluptuary world that will lift no vessels, that would bar all vessels.

At other times the civil engineer's structures appear to grow out of the earth with a natural grace that implies that fulfillment of an organic plan. Pierre Boulle, in *The Bridge Over the River Kwai,* writes: "An observer, blind to elementary detail but keen on general principles, might have regarded the development of the bridge as an uninterrupted process of natural growth." The bridge rose day by day, "majestically registering in all three dimensions the palpable shape of creation at the foot of these wild Siamese mountains . . . " Fifty years after the construction of the Eiffel Tower a Parisian recalled: "It appeared as if the tower was pushing itself upward by a supernatural force, like a tree growing beyond bounds yet steadily growing . . . Astonished Paris saw rising on its own grounds this new shape of a new adventure."

From the organic implications of the civil engineer's structures it is but a short step to the spiritual. Mighty works of concrete, steel, or stone, seeming alive but superhuman in scope, inevitably invoke thoughts of the divine. The ultimate material expressions of religious faith are, of course, the medieval cathedrals. They are usually defined as the material creations of religious men. But they can also be considered as magnificent works of engineering which, through their physical majesty and proportion, impel the viewer to think lofty thoughts. In *Mont Saint-Michel and Chartres* Henry Adams has conveyed a sense of the way in which these physical structures both reflect and evoke a spiritual concept:

> Every inch of material, up and down, from crypt to vault, from man to God, from the universe to the atom, had its task, giving support where support was needed, or weight where concentration was best, but always with the condition of showing conspicuously to the eye the great lines which led to unity and the curves which controlled divergence; so that, from the cross on the fleche and the keystone of the vault, down through the ribbed nervures, the columns, the windows, to the foundations of the flying buttresses far beyond the walls, one idea controlled every line.

William Golding, in his novel *The Spire,* has explored the theme of the interrelationship between construction and religion. Set in medieval England, the novel relates the story of the building of a cathedral tower, a tower which threatens to cause the collapse of the structure on which it rests. Priest and master builder confront each other, and the construction is accompanied by their dialogue, the dialogue between faith and technology. At one point the priest addresses the master builder in these words:

> My son. The building is a diagram of prayer; and our spire will be a diagram of the highest prayer of all. God revealed it to me in a vision, his unprofitable servant. He chose me. He chooses you, to fill the diagram with glass and iron and stone, since the children of men require a thing to look at. D'you think you can escape? You're not in my net . . . It's His. We can neither of us avoid this work. And there's another thing. I've begun to see how we can't understand it either, since each new foot reveals a new effect, a new purpose.

Not only cathedrals, but every great engineering work is an expression of motivation and of purpose which cannot be divorced from religious implications. This truth provides the engineer with what many would assert to be the ultimate existential experience.

I do not want to get carried away on this point. The age of cathedral building is long past. And, as I have already said, less than one-quarter of today's engineers are engaged in construction activities of any sort. But every man-made structure, no matter how mundane, has a little bit of cathedral in it, since man cannot help but transcend himself as soon as he begins to design and construct. As the priest of *The Spire* expresses it: "each new foot reveals a new effect, a new purpose."

In spite of the many ugly and tasteless structures that mar our cities and landscapes, public enthusiasm for building has survived relatively unscathed through the recent years of disenchantment with technology. The engineer, in company with architects, artists, and city planners, has kept alive the public faith in the potentiality for beauty, majesty, and spirituality in construction.

At a time when we are embarrassed to recall the grandiose pronouncements of so many of our predecessors, the First Proclamation of the Weimar Bauhaus, dating from 1919, retains its dignity and ability to inspire. It was the concept of architect Walter Gropius that great art in building grew out of craftsmanship, was in fact nothing other than craftsmanship inspired. His concept of craftsmanship included necessarily the essentials of civil engineering. "We must all turn to the crafts," he told his followers:

> Art is not a "profession." There is no essential difference between the artist and the craftsman. The artist is an exalted craftsman. In rare moments of inspiration, moments beyond the control of his will, the

grace of heaven may cause his work to blossom into art. *But proficiency in his craft is essential to every artist.* Therein lies a source of creative imagination. Let us create a *new guild of craftsmen,* without the class distinctions which raise an arrogant barrier between craftsman and artist. Together let us conceive and create the new building of the future, which . . . will rise one day toward heaven from the hands of a million workers like the crystal symbol of a new faith.

Enough, then, of the civil engineer and his wrestling with the elements, his love affair with nature, his yearning for immensity, his raising toward heaven the crystal symbol of a new faith. His existential bond to the earth, and expression of his own elemental being, need no further amplification, no additional testimonials.

Decision Making
in the Corps of Engineers:
The B. Everett Jordan Lake and Dam

P. AARNE VESILIND

A few years ago the U.S Army Corps of Engineers officers started wearing a big button on their uniforms. The highly visible non-regulation insignia proclaimed "THE CORPS CARES."

There is no doubt that the officers of the Corps care. They are people of high moral standard and take their responsibility to the public seriously. Why then are the environmental groups continually fighting the actions of the Corps? Why then did the late Justice William O. Douglas entitle a widely read essay on the Corps "The Public Be Damned"?[1] Why then is the Corps considered the diligent destroyer and not the protector of our environment?

This seeming incongruity is illustrated in this article describing the decision making with regard to the B. Everett Jordan Dam and Lake. As Samuel Florman suggests, it is indeed fun to build a dam. This enjoyment, coupled with the understandable desire to advance and be promoted within the Corps, is much stronger than the concern for the environment. The Corps may indeed care, but for what?

The Haw River, a major tributary of the Cape Fear River, cascades down the fall line above Fayetteville, North Carolina, and has been known to cause serious flooding in that eastern North Carolina city. The flood of 1945 was

An early version of this paper appeared in A. S. Gunn and P. A. Vesilind, *Environmental Ethics for Engineers* (Lewis Publishers, Ann Arbor, MI, 1982).
1.   Justice William O. Douglas, "The Corps of Engineers: The Public Be Damned," *Playboy*, July 1969.

particularly serious, causing over 2 million dollars of damage. Following a specific request by the people of Fayetteville, channeled through Senator Kerr Scott, the U.S. Army Corps of Engineers instituted a study of alternatives for flood protection.

The conventional solution to problems of flooding is to dam the offending river and thus capture the floodwaters behind the dam. In this case, however, the engineers encountered a problem. The Haw, at a point far enough upstream to assist Fayetteville, drops too rapidly and thus affords poor dam sites. The solution to this dilemma was ingenious: build a dam which captures most of the water in the Haw, but store it in a lake formed by a minor tributary, the New Hope Creek, forming a two-pronged lake. Following congressional approval, the construction of the New Hope Dam commenced in 1967.[2]

The first outspoken opponent of the project was Mr. Harold Cooley, member of Congress, who staked his reelection campaign on this issue. Although he was stoutly supported by the farmers of Chatham County who stood to lose prime farmland to the lake, he lost in the City of Raleigh, and was defeated. At that time, the most vigorous supporter of the dam was Senator B. Everett Jordan, who owned a textile mill on the Haw River, twenty miles above where the lake would reach. Senator Jordan's non-support of Mr. Cooley contributed to the election of Mr. James Garner, a Republican who supported the project and served for one term in Congress.

The next public opposition to the project emerged from North Carolina State University, where E. H. Weiser, a hydrologist, had calculated the flood probabilities and had concluded that the Corps of Engineers' flood damage projections were grossly inflated.[3]

Following the wide publication of this information, the North Carolina Conservation Council, a public interest organization, asked that the Corps prepare an Environmental Impact Statement (EIS) as required in the just-enacted National Environmental Policy Act. The Corps obliged by publishing an EIS in May, 1971, and a supplement was added in 1976. Although the EIS is theoretically supposed to be prepared as a planning document in order to judge the feasibility and wisdom of initiating a specific project, the writing of the EIS and subsequent court battles did little to hinder the clearing of the land and the dam construction.

Following the publication of the court-directed supplement to the EIS,

2. *New Hope Lake*, pamphlet prepared by the U.S. Army Corps of Engineers, Wilmington District, 1970.
3. E. Weiser, North Carolina State University, Raleigh, NC, private communication.

vigorous opposition developed to the project, based especially on the quality of water to be impounded. During this time, an independent benefit/cost study was conducted at the University of North Carolina by graduate students in environmental sciences and engineering. They found that a realistic calculation yielded a benefit/cost ratio of 0.3, where 1.0 would be required to justify the project.[4]

It was also argued convincingly, by nearly all of the expert water quality engineers and scientists at the University of North Carolina, North Carolina State University, and Duke University, that the lake would destroy thousands of acres of prime agricultural land, and that the water would be of questionable quality. Almost all of the experts agreed with this writer's conclusion that "if we looked for the absolute worst place to build a dam in North Carolina, we would not do much better than this site."[5]

Much of the water quality controversy centered on phosphorus. An analysis of the water which would flow into the lake showed that the level of nutrients was at least an order of magnitude higher than what would be necessary for accelerated eutrophication.[6]

The high residence time in the New Hope arm of the lake, coupled with its high nutrient loading and shallow depth, made the use of this water for recreation or other related uses highly questionable. Clearly phosphorus removal would be required by the towns of Durham and Chapel Hill, a cost not included in the Corps of Engineers benefit/cost analysis.

Responding to public pressure, the Corps decided to hire an independent disinterested consultant to establish once and for all the acceptability of the water quality. Hydrocomp, a respected hydrologic and water resources consulting firm from Palo Alto, California, was hired by the Corps to do the study.

The Hydrocomp report was published in 1976, as a supplement to the supplemental EIS, and showed conclusively that the water quality in much of the lake would be far below what the Corps had predicted, and that the New Hope arm of the lake would probably have serious water quality problems.[7]

4. "The New Hope Project – A Reevaluation," report by students at the Department of Environmental Sciences and Engineering, University of North Carolina in Chapel Hill, NC, 1974.
5. Transcript of trial, North Carolina Conservation Council vs. Corps of Engineers, U.S District Court, Greensboro, NC.
6. C. Weiss, "Water Quality in the Haw River and the New Hope Creek," Water Resources Research Institute, North Carolina State University, Raleigh, NC, 1973.
7. "Supplement to the Final Supplemental Environmental Impact Statement on the B. Everett Jordan Dam and Lake," Hydrocomp, Palo Alto, CA, 1976.

A decision was necessary. Should the Corps continue insisting, as before, that the lake was needed (as it clearly was not, based on the benefit/cost analysis), and that the water quality would be acceptable (as it would clearly not be, based on the study funded by the Corps), or should the wisdom of completing the project and filling the lake be reevaluated? Which of these options would be chosen?

At this point in the chronology, I will digress to describe the history, function, and operation of the Corps of Engineers in order to establish the framework in which this decision was made. I will then return to the question of the dam and show how the decision by the Corps is influenced by factors other than technical and economic considerations, or even the welfare of the people the Corps is supposed to serve.

The Corps of Engineers

The history of the U.S. Army Corps of Engineers stretches to the American Revolution, with the present Corps of Engineers tracing back to 1802 when it was formed by an act of Congress. Although originally meant to perform military duties, the Corps was instructed by Congress in 1824 to perform civilian duties as well, such as clearing snags from rivers. Since that time, the Corps' civilian duties have increased to projects totaling over 2.5 billion dollars per year.[8]

The Corps has an impeccable tradition, with a reputation as a dedicated and honest civil servant. It consists of a cadre of about 200 regular army officers who maintain a high morale among about 30,000 civilian employees. Typically, the career of a Corps officer begins at West Point, where only the brightest and best students receive appointments to the Corps. Assignments at various districts and at headquarters in Washington follow, with retirement as a colonel or general. The closeness and intimacy of this elite group of officers contributes greatly to the efficiency and effectiveness of the Corps. It is, in short, a select club of highly skilled professionals, beholden to each other and to the Corps.

The Corps of Engineers is, however, the only federal agency that doesn't follow the rules of executive/legislative conduct. Instead of incorporating its budget requests with the remainder of the administration plans, it

---

8. J. J. Lenny, *Caste System in the American Army: A Study of the Corps of Engineers and Their West Point System* (Greenberg Publishing, New York, 1949).

reports through the Office of Management and Budget directly to Congress, thus bypassing scrutiny by other agencies and even the President. The Secretary of the Army, who is the titular head of the Corps, has only limited power to interfere in the Corps' activities since the financial power comes from Congress.[9]

Every president since Franklin Roosevelt has tried, unsuccessfully, to either dismantle or curtail the powers of the Corps of Engineers. The Hoover Commission Task Force on Water Resources and Power recommended in 1949 that the civil functions of the Corps be transferred to the proposed Department of Natural Resources. No action was taken. In 1966, a bill sponsored by three senators to transfer some of the functions of the Corps died in committee. Even Lyndon Johnson, probably the best friend the Corps ever had, tried to keep it in line but failed. The last serious attempt to stifle the power of the Corps was by President Jimmy Carter when he recommended that over 40 Corps projects be abandoned as too costly and destructive to the environment. The political clout of the Corps was such, however, that only two were eventually stopped; the remainder continued as originally planned.

The Corps of Engineers has divided the United States into districts, usually based on river systems. A colonel is in command at each district, and the command is rotated every three years. The district commander is in direct charge of all of the projects in his district.

The Wilmington District office was, in 1976, headed by Colonel Homer Johnstone, who, following normal rotation, became the eighth engineer in charge of the New Hope Dam project, which by this time had been named in honor of its most ardent supporter, Senator B. Everett Jordan.

It was Colonel Johnstone who received the report from Hydrocomp which clearly showed that the dam had been a mistake. The construction of the dam was almost complete, however, and an admission that the project was indeed ill-considered might have significant repercussions. Colonel Johnstone had to make a decision: Should he proceed with the project and eventually fill the lake, or propose some other alternative?

Possible Options Concerning the B. Everett Jordan Dam

1. Stop all further construction and abandon the project.

---

9. G. Laycock, *The Diligent Destroyers* (Audubon/Ballantine, New York, 1970).

2. Finish the project but leave the lake-bed dry. This would provide for maximum flood protection downriver, since the largest possible volume of water could be retained during floods, but benefits such as recreation and water supply would not be realized. This alternative would also save money, since some of the road construction necessary to elevate roads to above flood level (construction which extended well into the 1980s) would not be necessary.

3. Finish the project as planned and fill the lake.

Although all disinterested expert opinion and formal testimony counseled against filling the lake, and choosing alternative 2, such a choice would in effect be an admission that the dam was a mistake. And if Colonel Johnstone admitted that the Corps had made a mistake, he would be indirectly criticizing his 7 predecessors (some of whom may have by then been generals in the Pentagon). What would that do to his career, and to the image of the Corps? Thus the decision became more than a technical or economical decision. Given the spirit of the Corps, the close-knit camaraderie and "old-boy" method of promotion, deciding that seven of one's predecessors (and presumably superiors) were wrong in their analysis of this project would have been professional suicide. Further, the image and reputation of the Corps as a monolithic and technically infallible organization would have been compromised. And finally, what would a decision to alter the project have done to the many politicians who had stoutly supported the original decision? It would have implied that they had bet on the wrong horse, and they would have concluded that the Corps had let them down – a notion which might have been reflected in the appropriations bill working its way through Congress.

What Really Happened?

On the day the Hydrocomp report was released to the public, the Corps accompanied it with a press release which had as its headline "WATER QUALITY IN JORDAN LAKE TO BE BETTER THAN ANTICIPATED."[10] The story went on to say that although the report showed that much of the lake would probably not be useful for recreation and water supply, it would not be a "cesspool." The

---

10. *The Chapel Hill Newspaper*, Chapel Hill, NC.

straw man tactic worked; the newspapers printed the headline verbatim, and the lake is filled. At the present time [1997] the towns of Chapel Hill and Durham are spending millions of dollars for phosphate removal processes at their wastewater treatment plants.

In all fairness to Colonel Johnstone, he may in fact have concluded, even in the face of overwhelming technical and public opinion, that the just and proper course of action was to continue with the project. We will never know what influenced the Colonel's decision. What we *do* know is that the lake was filled and that Colonel Johnstone went from Wilmington to Korea and became a brigadier general before being honorably retired from the Corps of Engineers.

Future Generations
and the Social Contract

KRISTIN SHRADER–FRECHETTE

Engineers, while having fun, often do so at the expense not only of present people but of future people. There are those who insist that we owe nothing to our progeny, because, after all, what have they ever done for us? But does it make sense to sacrifice some benefits today for the good of these future people?

Kristin Shrader-Frechette is a leading environmental ethicist, having written numerous articles and several books on the responsibilities we ought to have to environmental quality. This is an excerpt from her book, *Environmental Ethics*, on the rights of future generations.

When one raises the question of the rights of future generations, at least three important issues come to mind. First, is there an ethical framework within which the existence of such rights can be rationally substantiated? Secondly, assuming these rights exist, is it possible either to calculate or to meet the legitimate interests of all persons, present and future? Thirdly, would recognition of the rights of future generations diminish the extent to which the needs of current individuals (especially the poor or socially disenfranchised) were served?

In the following pages I will discuss all three of these questions, although my emphasis will be on the first one, since it is both logically and chronologically prior to the other two. Moreover, although there are numerous ethical frameworks (e.g., altruism, natural law theory rule- or act-utilitarianism) within which the question of the rights of future persons may

Excerpted from K. S. Shrader-Frechette, *Environmental Ethics* (The Boxwood Press, Pacific Grove, CA, 1981). Used with permission.

be investigated, I will analyze primarily the social contract basis for affirming such rights. The main reasons for discussing the contractarian view are that it is easily understood and commonsensical, it has a long tradition in the history of ethics, and most importantly, it is the camp from which many of the major attacks on the existence of the rights of future persons have issued. Hence, an important parameter for determining whether present individuals have certain obligations, because of the rights of future persons, is the extent to which current and forthcoming generations share a social contract. Are both members of the same "moral community"?

According to M. P. Golding, who wrote one of the first and major analyses of the rights of future generations, a moral community may be constituted in one of two ways. It may be initiated either by "an explicit contract between its members" or by "a social arrangement in which each member derives benefits from the efforts of the other members."[1] Since it is impossible for present persons to enter into an explicit legal bargain with future individuals, several authors have maintained that social contracts between current and forthcoming persons are a chronological impossibility; moreover, they argue, the second type of moral community is also impossible, since present individuals can benefit future ones, but not vice versa. Because remote generations can share neither type of moral community with us, they reason, members of distant posterity cannot be said to have rights based on a social contract or arrangement.[2]

Some authors maintain that, even though future persons do not have rights based either on an explicit contract or on a reciprocal agreement, they might have rights equal to ours, provided that they share the same interests or social ideal as we. In other words, if our remote descendants share the same conception as we for the good life of humans, then altruism and fellow-feeling urge us to affirm that they are members of the same moral community, and therefore have the same rights as we. In his classic article, Golding argued that, as a result of the complexity and rapidity of technological change, it is impossible to know the conditions of life of our remote descendants. Because it is impossible to foresee whether they will maintain the same conception of the good life as we, Golding concluded that they could not be said to share our social ideal and hence the same rights as we.

Interesting as is the case against the existence of rights for future persons, it suggests investigation of at least three questions whose answers could lead

1. M. P. Golding, "Obligations to Future Generations," *The Monist*, vol. 56, no. 1 (1972).
2. J. B. Stearns, "Ecology and the Indefinite Unborn," *The Monist*, vol. 56, no. 4 (1972).

one to affirm that members of future generations do have rights. First, is intergenerational reciprocity impossible, as Golding and others say it is? Secondly, is explicit reciprocity a necessary condition for all social contracts based on self-interest? And thirdly, is it plausible to reason that we do not know whether future persons will have the same conception of the good life, and therefore that they cannot be said to have rights bearing on our present obligations? Let us examine these questions in sequence.

There are several considerations suggesting that intergenerational reciprocity might be possible and future generations might be said, indirectly, to benefit us in exchange for our recognition of their alleged rights. The first type of reciprocity has been described by social scientist Walter Wagner, who argues that if we recognize the rights of future persons, we obtain in return a greater degree of happiness and self-actualization. Active concern for the interests of our remote descendants, says Wagner, increases our empathy and our compassion and thus benefits us as individuals.[3] Hence, indirectly, future persons might be said to have a reciprocal relationship with some current individuals.

A more frequently discussed, and perhaps more important, case of intergenerational reciprocity is based on the notion that since our forebears have benefitted us in numerous ways, we have obligations to aid remote posterity. In other words, the presupposition of this social contract is not that generation A benefits generation B and vice versa, but that A benefits B, B benefits C, C benefits D, and so on. One writer has cited the rationale behind this ancient notion of intergenerational dependence by quoting an old manuscript of Benedictine monks, "On the Conservation of Pine Forests":

> . . . no one who plants a fir-tree can hope to fell it when it is fully grown, no matter how youthful the person is. In spite of this the most *sacred obligation* is to replant and husband these pine forests. If we sweat for the benefit of posterity, we should not complain as we reap the results of the efforts of our forefathers.[4]

Or, as Peter Faulkner put it: "This generation owes to posterity concern of the same quality and degree that [our ancestors] . . . devoted to all generations following theirs and that made our present happiness possible."[5]

---

3. W. C. Wagner, "Future Morality," *The Futurist*, vol. 5, no. 5 (1971).
4. V. R. Potter, "Evolving Ethical Concepts," *BioScience*, vol. 27, no. 4 (1977).
5. P. Faulkner, "Protection for Future Generations," *Frontiers*, vol. 42, no. 4 (1978).

This particular concept of intergenerational reciprocity appears to have been even more explicitly formulated in terms of the Japanese concept of *on*, whose meaning is close to the Western notion of "obligation"; one author, discussing *on*, wrote: "One makes past payment *on* to one's ancestors by giving equally good or better" things to posterity. Thus the obligations one has to past generations "are merely subsumed under *on*" to future generations. In other words, future persons may be thought of as proxies for past persons to whom we are indebted; we repay these debts by doing for our descendants what our ancestors have done for us.

If one accepts either Wagner's, Faulkner's or the Japanese notion of intergenerational reciprocity, then perhaps there are grounds for affirming that present and future persons share a social contract and hence some bases for affirming that future generations may be said to have rights. Even if one does not accept their reasoning, however, one might still have reasons for believing that social contracts can exist between present and future generations. This brings us to a second question raised by the case against the rights of remote descendants. Is explicit reciprocity a necessary condition, as Golding and others say it is, for all social contracts based on self-interest? There are a number of considerations suggesting that reciprocity is not necessary for all moral communities. John Rawls, for example, author of perhaps the most widely discussed current ethical theory, proposes a "thought experiment" which shows that the social contract is not based on reciprocity but on moral reasoning. Although Rawls' account is somewhat complex and hypothetical, the ethical reasoning he advocates is based on a familiar insight (viz., the duty to base one's contracts on the golden rule, on walking in the other person's shoes).

According to his view, one may think of a purely hypothetical situation (1) in which each of us seeks to protect our possible interests but (2) in which none of us knows his particular place in society, natural ability, assets, or liabilities. Rawls calls a situation in which (1) and (2) are the case, "the original position." If one were in "the original position," he says, one would formulate a social contract based both on reason and on self-interest. Rawls says this is because:

> . . . principles of justice are chosen behind a veil of ignorance. This ensures that no one is advantaged or disadvantaged in the choice of principles by the outcome of natural chance or the contingency of social circumstances. Since all are similarly situated and no one is able

to design principles to favor his particular condition, the principles of
justice are the result of a fair agreement or bargain.[6]

Thus, if one thinks of all humans as being in "the original position," it is
clear, provided that they are capable of rational deliberation, that they all
would agree on the same set of moral principles based on equity. For example, if all members of the human race were in the hypothetical "original
position," then no one would know of which generation he were a member.
Because of this ignorance, the only reasonable moral principle for any person to follow would be that members of all generations ought to have equal
rights. As Rawls puts it, no rational person would agree to a contract that
would allow "an enduring loss for himself in order to bring about a greater
net balance of satisfaction"; hence, all persons would choose to follow principles of "equality in the assignment of basic rights and duties." Thus if one
conceives of an alternative account of social contracts, not based on reciprocity but instead on rationality, self-interest, and justice, then one has a
contractarian foundation for rights of future generations.

Less abstract and hypothetical than Rawls' account, Daniel Callahan's
view of the social contract also does not involve notions of reciprocity. His
example is of the parent–child relationship; it deserves special treatment
here, not only because it might provide a counter-example to Golding's
claim, but also because it might serve as a model for understanding the
rights of present and future persons.

According to Callahan, certain types of contracts come into being, not
because of any agreement based on prearranged reciprocal benefits, but simply because one party to the contract chooses to accept an obligation. Children are not asked whether they wish to be born, says Callahan, but the fact
that parents take upon themselves an obligation toward them means that a
contract between the two has been initiated. Anyone who values his life, he
says, has thereby accepted an implicit obligation to his mother and father.
This partnership or contract exists, says Callahan, because children are in
debt to their parents.[7]

In describing the parent–child relationship, Callahan has provided an
example of a social contract with at least two distinctive features. First, the
children are not asked if they wish to be a part of the agreement. Secondly,

---

6. J. Rawls, *A Theory of Justice* (Harvard University Press, Cambridge, MA, 1971).
7. D. Callahan, "What Obligations Do We Have to Future Generations?" *The American Ecclesiastical
    Review*, vol. 164, no. 4 (1971).

the parents' obligation is not contingent on the fact that the child reciprocates what has been done for him, since presumably they would still have the duty to care for their offspring even if he or she were unable or unwilling to reciprocate in any way. Hence, if one accepts Callahan's apparent assumption, that an implicit contract may be initiated when one person is knowingly or unknowingly put in debt to another, then he has an example of a contract not necessarily involving notions of reciprocity. As Callahan suggests, this model may provide a prototype for relationships between present and future generations, since the latter (like offspring) are not asked whether they wish to enter an agreement and either cannot or may not return the benefits given them. Moreover, present generations (like parents) may not always know what future generations (like children) need. Just as lack of this complete knowledge does not remove parental obligations, so also it could be argued that it does not eliminate duties to the future.[8]

What of the contention that, because we do not know whether future persons will have our same conception of the good life, therefore they cannot be said to have rights bearing on our present obligations? On the one hand, it appears plausible to affirm that we do not know what to desire for the future. For example, we do not even know whether we have an obligation to future generations to maintain current genetic quality, because a particular "chromosome is 'deleterious' or 'advantageous' only relative to given circumstances," and we do not know the conditions of the future.[9] On the other hand, however, it seems that members of future generations do have a number of "proxies" to speak now about their conception of the good life. Although we proxies are uncertain about whether the good life in the future will include use of great quantities of fossil fuel, for example, we do know, says Joel Feinberg, that future persons "will have an interest in living space, fertile soil, fresh air, and the like."[10] Hence, he concludes, "Whoever these human beings may turn out to be, and whatever they might reasonably be expected to be like, they will have interests that we can affect, for better or worse right now . . . The fact of their interest-ownership is crystal clear, and that is all that is necessary to certify the coherence of present talk about their rights."[11] In other words, Feinberg claims that the specific nature of the

8. See ibid.
9. See Golding, "Obligations."
10. J. Feinberg, "The Rights of Animals and Unborn Generations," in T. A. Mappes and J. S. Zembaty, eds., *Social Ethics: Morality and Social Policy* (McGraw-Hill, New York, 1977).
11. *Ibid.*

interests of future generations is not significant, but only that they have interests. Hence he believes that knowing the exact nature of our descendants' notion of the good life is not necessary for awarding them rights; rather, all that is required is our knowledge that they will have a conception of the social ideal which we can affect.

Feinberg's answer, however, does not resolve the difficulty raised by those who deny that future persons have rights. What good does it do to affirm that future generations have interests and rights which we can affect, and therefore that we have a duty to safeguard them, if we do not know, specifically, what we are obligated to protect? Also, if we assume, with Feinberg, that future persons will have an interest in "living space, fertile soil, fresh air, and the like," then we will not have answered the critical question but only begged it. In other words, we will not have said how or why we know what the interests of distant generations are, but we will have taken it for granted that some of their needs will be the same as ours. Given rapid technological change and our ability to compensate for certain environmental mistakes, this might be a false assumption.

The real error of the argument Feinberg attempts to refute, however, is the presupposition that, because we are in ignorance about the specific welfare of future generations, we are therefore justified in assuming that their interests will not be the same as ours. This presupposition comes down to commission of the fallacy of *argumentum ad ignorantiam*, the argument from ignorance. From the premise that we are ignorant about *x*, it never follows that *x* is or is not of a particular nature, or that we are justified in assuming that *x* is or is not of a particular nature. From such a premise, nothing follows. On the same basis, our ignorance about the specific welfare of future generations would not logically justify our concluding that those interests are the same as ours. Where, then, does all this leave us? Is it accurate to assume that, because we do not know whether future persons will have our same conception of the good life, therefore they do not belong to the same moral community and consequently they cannot be said to have rights?

There are several reasons for believing that our ignorance in this regard does not enable us to affirm that future generations have no rights. Although it is true that we do not know what, positively, is in the interests of our descendants, we *do* have some idea of what might be very dangerous to them. We know, for example, of the long-lived toxicity of both plutonium and DDT. Since both hazards have harmful mutagenic and carcinogenic consequences, it is implausible to assume that increasing the permanent contamination of the planet by these substances will not adversely affect the interests of future generations. Thus, in at least some circumstances, present knowledge of harmful, long-term consequences might be sufficient to enable us to claim that future generations have an interest in not being subjected to higher incidence of cancer and genetic damage. And if they have such an interest in common with us, then it is conceivable that, at least in this sense, we are members of the same moral community and have equal rights to protection against similar damages.

In the same vein, although the particulars of some future conception of the good life might not be known to us now, on the basis of Rawls' thought experiment, we do know that any person (past, present, or future) is likely to desire an ethical code based on equity. This means, for example, one which does not permit treating humans as means, leaving one's debts to others to pay, distributing resources inequitably, ignoring due process, or failing to protect the utterly helpless members of society. In other words, we still have a great deal of information about what we ought not to do, even though we do not (specifically) know what to do on behalf of later generations. If present persons pollute the atmosphere with growing amounts of permanent carcinogenic and mutagenic materials, then it is difficult to see how this behavior would not constitute both using future generations as means for the ends of present generations, as well as leaving our debts for our descendants to pay.[12] Obviously, people in Rawls' "original position" would condemn such an action. Likewise, if present individuals were to use the lion's share of nonrenewable resources for which there were no known substitutes, then this would be a clear violation of principles of distributive justice which everyone would condemn, were they in "the original position." Similarly, since it will be impossible for future persons to collect damages for what current individuals have inflicted on them (e.g., permanent carcinogens), then it is likely that a great many actions of this generation constitute imposing a degraded environment on posterity without due process, a constitutional right. Because of the impossibility of providing complete due process to all our descendants, they are, in at least one sense, in a position of utter helplessness. And, as Hans Jonas points out, "Utter helplessness demands utter protection"; this is an "inflexible principle."[13] Thus if we are in ignorance regarding the social ideal of future generations, we nevertheless have an

12. J. L. Huffman, "Individual Liberty and Environmental Regulation," *Environmental Law*, vol. 7, no. 3 (1977).

13. H. Jonas, "Philosophical Reflections on Experimenting with Human Subjects," in K. J. and R. R. Struhl, eds., *Ethics in Perspective* (Random House, New York, 1975).

obligation to provide them with "utter protection." This is another way of saying that, since future generations are helpless to change what we force on them, it is incumbent on us to impose lesser risks on them than those we would voluntarily take on ourselves, since they must accept such hazards involuntarily. As William Lowrance put it, "Risks that citizens have to bear involuntarily are the ones the government has the greatest responsibility to regulate."[14]

There is also another reason for arguing that, even if we are in ignorance about the social ideal of our remote descendants, this is not sufficient grounds for concluding that their possible interests ought to be ignored. If we admit that since we do not know "the good life" for future generations, therefore we cannot "plan for them" by giving them certain rights, then by virtue of consistency, it is also incumbent on us not to plan for them in a different way. In other words, if one follows the logic of those who deny that future persons have rights, then one is also bound not to destroy the possibility of distant generations' planning for themselves. Obviously certain of our acts in the present could preempt the choices of future generations. But such preemption would be morally unjustified if we did not know whether we were destroying the possibility for future persons' attaining their social ideal, whatever it might be. Thus even if we admit our ignorance about the interests of later generations, they have rights not to have damages inflicted upon them, and rights not to have their choice of "the good life" preempted.

A final consideration is that, since we cannot know the social ideal of future generations, the morally safe course to follow is to assume it will not be different from our own. If we do something in the present which would be extremely hazardous to persons in the future, if they shared our conception of the good life, then it would be morally irresponsible to assume that our behavior is totally acceptable. It would not be unlike using a gun to shoot at moving underbrush, before deciding whether the source of the motion were our hunting partner or a game animal. In other words, when a situation of ignorance exists, the morally responsible course is to follow the position least likely to violate possible rights. As Callahan says, since we cannot know what the social ideal will be for future generations, "we should act on the assumption that it will not be all that dissimilar from our own."

14. W. W. Lowrance, *Of Acceptable Risk: Science and the Determination of Safety* (William Kaufman, Los Altos, CA, 1976).

Consumption, Conservation, Use, and Preservation

ALASTAIR S. GUNN

One of the problems with the development of environmental ethics in engineering is the lack of definitions. In this article, four of the most confused and confusing terms – consumption, conservation, use, and preservation – are defined and discussed.

Humans need air, land, water, energy, materials, and their products to live, collectively termed *resources,* and it can be assumed that individuals have the right to access these resources in order to ensure at least a minimally decent life. What standard is "minimally decent" is not discussed here.

Total human depletion of resources is a function of three variables: *nature of resource* (e.g., renewable vs. nonrenewable), *per capita access,* and *population size.* The effect on the total resource may be *direct* (e.g., burning gasoline in cars) or *indirect* (e.g., destroying forests by acid precipitation) and it may be *intentional* (e.g., switching on the spa pool), *unnoticed* (e.g., lead poisoning in ancient Rome), *misinterpreted* (e.g., environmental effects of PCB contamination mistakenly believed to be due to DDT), *recognized as a fact but not as a problem* (e.g., faith in the technological fix), *covered up* (e.g., contrary-to-fact assertions by U.S. electricity companies that coal-burning power stations do not produce acid precipitation), etc.

Human accessing of resources may be represented as lying on a continuum, four areas of which we label *consumption, conservation, use,* and *preservation.*

Consumption is the using up of resources so that they become lost forever, or for the reasonably foreseeable future.

An apple is consumed if it is eaten. The energy value of a quantity of oil

Report for the New Zealand Environmental Council, 1984.

or gas is consumed if it is burned. The productive capacity of a person, animal, or area of land is consumed if it is worked to death (e.g., concentration camp slave labor, stagecoach horses, overgrazed and overploughed land.) The so-called cowboy or frontier ethic[1] consumes resources profligately either not knowing or not caring that all resources are finite and most resources are nonrenewable or renewable only at a slower rate than they are being consumed.

So long as a resource is effectively inexhaustible and our uses do not degrade it we may consume it as we please. For instance, we can all breathe deeply without using up the oxygen in the air providing we do not overproduce carbon dioxide and/or destroy the plants which convert the CO_2 back into oxygen. Breathing uses up some oxygen but it does not use up *the* oxygen because more will continue to be produced, and in any case the resource is so immense as to be practically unlimited. Fossil fuel reserves are fixed, however, and are being depleted at rapid rates, while few large reserves remain to be discovered.

Each use of coal, oil, or natural gas also consumes part of the total resource. That the resource will eventually be consumed has nothing whatever to do with the rate at which it is depleted or the manner of use. There is no way to continue to consume *any* gas without eventually consuming the total resource, and no way of guaranteeing its availability to all future generations.

Conservation is the consumption of resources at a rate perceived to be slow.

Individual quanta of a renewable resource can be consumed indefinitely provided that the depletion rate does not exceed the rate of renewal, for instance the concept of a sustained yield from a fishery. Excessive depletion, or harvesting, or "cropping," consumes the resource, as when a species is exterminated by overhunting. Conservationist draw-off consumes only the net yield of a resource without consuming the resource itself, as when one spends only the interest on an investment, leaving the capital intact.

Nonrenewable resources, as just explained, do not have a net yield, so that consumption of any of the resource depletes all of it. Conservation of fossil fuels, therefore, can only take the form of depleting at a lower rate in order to leave a larger total resource to be consumed in the future. It is possible to leave more of a resource for future generations by consuming less per capita, by consuming the resource more efficiently (which comes to the same thing), or

1.   K. S. Schrader-Frechette, "Frontier Ethics," in *Environmental Ethics* (Boxwood Press, Pacific Grove, CA, 1981).

by having fewer consumers. The slower the rate of depletion, the more time will be available for future generations to develop renewable alternatives.

Preservation is the non-use of resources, or rather the attitude and associated behavior that certain natural objects, features, and systems are not for human use. In its "purest" form it is the protection of natural objects (etc.) from any interventions with humans, for instance, a perception of wilderness as an area in which humans have no place and the passing and enforcement of legislation to exclude humans.[2] This attitude is exemplified by those who oppose all forms of space exploration on the grounds that any human intrusion into space and the planets and stars therein is unwarranted and a form of pollution. In practice something close to this is indicated by the "hands-off" attitude to Antarctica of many environmental groups, or the closing of extremely fragile systems such as Sunset Crater in Arizona, a geological feature of such fragility that not even occasional scientific teams are allowed to tread on it because every footstep erodes its surface.

To preserve that which is nonrenewable, then, is to leave it alone. To preserve a natural gas field or a coal seam is to refrain from removing any of it – "leaving it in the ground."

It is not clear whether this concept of use can be applied to all uses of renewable resources. Consider, for example, the hunting of wildlife. In many countries, many native species of birds and mammals are given varying degrees of protection – sometimes absolute protection – from direct killing or injury, though often their habitat may be legally destroyed, for instance by clear-cutting forests. Some species and populations are conserved, by permitting only a limited hunting season. Other species enjoy no legal protection, while yet others, usually introduced species, are viewed as pests to be controlled or even exterminated. Responsible hunters consume individual animals but they seek to protect the species, or a particular population, in order that there will be individual animals in the future to be consumed. Hunting is not use, because the successfully hunted animals are not returned to the wild. Sports fishing, likewise, usually consumes individual fish, though anglers who release trout and other game fish unharmed may be said to use rather than consume the animals.

Few people appear to take a preservationist view of resources such as natural gas, or of any other inanimate energy resources. All the views on the subject seem to agree that the resource should be consumed. The arguments

2. Linda Graber, *Wilderness as Sacred Space* (American Society of Geographers, Washington, DC, 1980).

are only about the rate and manner of depletion of current fields and, presumably, any other fields that may be discovered. There seems to be little point in discussing the preservationist plan of leaving the remaining gas where it is forever, even if that is technologically feasible in the case of natural gas.

My concept of *use* departs from the ordinary language concept: it is here defined as the employment of at least part of a resource by humans and its ultimate return to something very close to its original state. Some uses have no effect, for instance, filming a wild animal without its knowledge and without in any way altering the ecosystem of which it is a part. At the other extreme, sophisticated water use systems extract natural water from a river, use it for a variety of domestic, commercial, and industrial purposes, purify it, and return it to the same river as pure water.

Temporary disturbances of an ecosystem may be seen in this light too, for instance, a careless bird-watcher frightens away a feeding or nesting bird, which returns when the ornithologist departs: she has used the bird for scientific or recreational purposes but the situation is essentially the same shortly after the use as it was shortly before.

Claims by miners to be able to reclaim or restore land after strip-mining are claims that the land will be used rather than consumed. This is not really what happens, however, because the minerals have been removed. Mining (this is not a value judgment) is not use but consumption, unless only the replaceable top few meters ("overburden") of a landform are considered.

Clear-cutting of a native forest and replanting with a different species is obviously not use, since the forest is replaced with different flora and fauna. What about clear-cutting and replanting with the original species? Suppose an exhaustive inventory of the original forest were taken and an exact replica were planted? Some philosophers have taken the view that second-growth forest is importantly different from "original" forest, in just the same way as a perfect copy of a famous work of art differs from the original: it lacks the history of the original.[3] If you admire Andrew Wyeth paintings, someone who wants to please you on your birthday, being unable to afford a real Wyeth, might commission an expert forger to produce a thoroughly convincing fake. You would probably be delighted with your gift. But what if you learned the truth, that your "Wyeth" was a fake? Your disappointment wouldn't be based on aesthetic grounds since the picture would still look the

---

3.   Robert Elliott, "Faking Nature," *Inquiry,* vol. 25 (1982); Alastair S. Gunn, "The Restoration of Species and Natural Environments," *Environmental Ethics,* vol. 13 (1991).

same. But it wouldn't be a real Wyeth: it would have a different history, origin, provenance.

Maybe, then, use doesn't necessarily leave us (or nature) as well off as we (or it) began. Maybe wild rivers lose something if somewhere on the way part of the water is drawn off, cycled through a piggery and tertiary treatment system, and pumped back into the river, even if the discharge is chemically and biologically indistinguishable from the intake water.

The Tragedy of the Commons

GARRETT HARDIN

Garrett Hardin (1915–) has been called "one of the intellectual leaders of our time" by *Science* magazine. Yet he probably would not think of himself as that, because he has been, since the tumultuous 1960s, preaching what he considers simple good sense. The tenor of all of his writings is: "Okay, now what is the bottom line? What is the unemotional, unvarnished truth in this matter? I don't care who it offends as long as it is the truth." Understandably, he has been criticized and even vilified by numerous writers who decry his inhumanity and insensitivity. But the opposite is true. Hardin writes *about* humanity and all of our problems and foibles. While he may at times appear crass, he is always thought-provoking. He shakes the trees to see what might drop and on whose head.

In this piece he starts by discussing the nature of what he calls "the tragedy of the commons," and in the remainder of the article (not reproduced here) he applies this concept to the problem of overpopulation and birth control. The "tragedy" applies equally well to many other environmental concerns and seems to be at the very root of our problems in achieving a stable, livable environment.

In economic affairs, *The Wealth of Nations* (1776) popularized the "invisible hand," the idea that an individual who "intends only his own gain," is, as it were, "led by an invisible hand to promote . . . the public interest." Adam Smith did not assert that this was invariably true, and perhaps neither did any of his followers. But he contributed to a dominant tendency of thought that has ever since interfered with positive action based on rational analysis, namely, the tendency to assume that decisions reached individually will, in fact, be the best decisions for an entire society. If this assumption is correct it justifies the continuance of our present policy of laissez-faire in repro- duction. If it is correct we can assume that men will control their individual fecundity so as to produce the optimum population. If the assumption is not correct, we need to reexamine our individual freedoms to see which ones are defensible.

Tragedy of Freedom in a Commons

The rebuttal to the invisible hand in population control is to be found in a scenario first sketched in a little-known pamphlet in 1833 by a mathematical amateur named William Forster Lloyd (1794–1852). We may well call it "the tragedy of the commons," using the word "tragedy" as the philosopher Whitehead used it: "The essence of dramatic tragedy is not unhappiness. It resides in the solemnity of the remorseless working of things." He then goes on to say, "This inevitableness of destiny can only be illustrated in terms of human life by incidents which in fact involve unhappiness. For it is only by them that the futility of escape can be made evident in the drama."

The tragedy of the commons develops in this way. Picture a pasture open to all. It is to be expected that each herdsman will try to keep as many cattle as possible on the commons. Such an arrangement may work reasonably sat- isfactorily for centuries because tribal wars, poaching, and disease keep the numbers of both man and beast well below the carrying capacity of the land. Finally, however, comes the day of reckoning, that is, the day when the long- desired goal of social stability becomes a reality. At this point, the inherent logic of the commons remorselessly generates tragedy.

As a rational being, each herdsman seeks to maximize his gain. Explicitly or implicitly, more or less consciously, he asks, "What is the utility *to me* of adding one more animal to my herd?" This utility has one negative and one positive component.

1) The positive component is a function of the increment of one animal. Since the herdsman receives all the proceeds from the sale of the addi- tional animal, the positive utility is nearly +1.
2) The negative component is a function of the additional overgrazing cre- ated by one more animal. Since, however, the effects of overgrazing are

shared by all the herdsmen, the negative utility for any particular decision-making herdsman is only a fraction of −1.

Adding together the component partial utilities, the rational herdsman
concludes that the only sensible course for him to pursue is to add another
animal to his herd. And another; and another . . . But this is the conclusion
reached by each and every rational herdsman sharing a commons. Therein
is the tragedy. Each man is locked into a system that compels him to increase
his herd without limit − in a world that is limited. Ruin is the destination
toward which all men rush, each pursuing his own best interest in a society
that believes in the freedom of the commons. Freedom in a commons brings
ruin to all.

Some would say that this is a platitude. Would that it were! In a sense, it
was learned thousands of years ago, but natural selection favors the forces of
psychological denial. The individual benefits as an individual from his ability to deny the truth even though society as a whole, of which he is a part,
suffers. Education can counteract the natural tendency to do the wrong
thing, but the inexorable succession of generations requires that the basis for
this knowledge be constantly refreshed.

A simple incident that occurred a few years ago in Leominster, Massachusetts, shows how perishable the knowledge is. During Christmas shop-
ping season the parking meters downtown were covered with plastic bags
that bore tags reading: "Do not open until after Christmas. Free parking
courtesy of the mayor and city council." In other words, facing the prospect
of an increased demand for already scarce space, the city fathers reinstituted
the system of the commons. (Cynically, we suggest that they gained more
votes than they lost by this retrogressive act.)

In an approximate way, the logic of the commons has been understood for
a long time, perhaps since the discovery of agriculture or the invention of
private property in real estate. But it is understood mostly in special cases
which are not sufficiently generalized. Even at this late date, cattlemen leas-
ing national land on the western ranges demonstrate no more than an
ambivalent understanding, in constantly pressuring federal authorities to
increase the head count to the point where overgrazing produces erosion
and weed-dominance. Likewise, the oceans of the world continue to suffer
from the survival of the philosophy of the commons. Maritime nations still
respond automatically to the shibboleth of the "freedom of the seas." Pro-
fessing to believe in the "inexhaustible resources of the oceans," they bring
species after species of fish and whales closer to extinction.

The National Parks present an instance of the working out of the tragedy of the commons. At present, they are open to all, without limit. The parks themselves are limited in extent – there is only one Yosemite Valley – whereas population seems to grow without limit. The values that visitors seek in the parks are steadily eroded. Plainly, we must soon cease to treat the parks as commons or they will be of no value to anyone.

What shall we do? We have several options. We might sell them off as private property. We might keep them as public property, but allocate the right to enter them. The allocation might be on the basis of wealth, by use of an auction system. It might be on the basis of merit, as defined by some agreed-upon standards. It might be by lottery. Or it might be on a first-come, first-served basis, administered to long queues. These, I think are all the reasonable possibilities. They are all objectionable. But we must choose – or acquiesce in the destruction of the commons that we call our National Parks.

Pollution

In a reverse way, the tragedy of the commons reappears in problems of pollution. Here it is not a question of taking something out of the commons, but of putting something in – sewage, or chemical, radioactive, and heat wastes into water; noxious and dangerous fumes into the air; and distracting and unpleasant advertising signs into the line of sight. The calculations of utility are much the same as before. The rational man finds that his share of the cost of the wastes he discharges into the commons is less than the cost of purifying his wastes before releasing them. Since this is true for everyone, we are locked into a system of "fouling our own nest," so long as we behave only as independent, rational, free-enterprisers.

The tragedy of the commons as a food basket is averted by private property, or something formally like it. But the air and waters surrounding us cannot readily be fenced, and so the tragedy of the commons as a cesspool must be prevented by different means, like coercive laws or taxing devices that make it cheaper for the polluter to treat his pollutants than to discharge them untreated. We have not progressed as far with the solution of this problem as we have with the first. Indeed, our particular concept of private property, which deters us from exhausting the positive resources of the earth, favors pollution. The owner of a factory on the bank of a stream – whose property extends to the middle of the stream – often has difficulty seeing why it is not his natural right to muddy the waters flowing past his door. The law, always behind the times, requires elaborate stitching and fitting to adapt it to this newly perceived aspect of the commons.

The pollution problem is a consequence of population. It did not much matter how a lonely American frontiersman disposed of his waste. "Flowing water purifies itself every 10 miles," grandfather used to say, and the myth was near enough to the truth when he was a boy, for there were not too many people. But as population became denser, the natural chemical and biological processes became overloaded, calling for a redefinition of property rights.

How to Legislate Temperance?

Analysis of the pollution problem as a function of population density uncovers a not generally recognized principle of morality, namely: *the morality of an act as a function of the state of the system at the time it is performed.* Using the commons as a cesspool does not harm the general public under frontier conditions, because there is no public; the same behavior in a metropolis is unbearable. A hundred and fifty years ago a plainsman could kill an American bison, cut out only the tongue for his dinner, and discard the rest of the animal. He was not in an important sense being wasteful. Today, with only a few thousand bison left, we would be appalled at such behavior.

In passing, it is worth noting that the morality of an act cannot be determined from a photograph. One does not know whether a man killing an elephant or setting a fire to the grassland is harming others until one knows the total system in which his act appears. "One picture is worth a thousand words," said an ancient Chinese; but it may take 10,000 words to validate it. It is as tempting to ecologists as it is to reformers in general to try to persuade others by way of the photographic shortcut. But the essence of an argument cannot be photographed; it must be presented rationally – in words.

That morality is system-sensitive escaped the attention of most codifiers of ethics in the past. "Thou shalt not . . . " is the form of traditional ethical directives which make no allowance for particular circumstances. The laws of our society follow the pattern of ancient ethics, and therefore are poorly suited to governing a complex, crowded, changeable world. Our epicyclic solution is to augment statutory law with administrative law. Since it is practically impossible to spell out all the conditions under which it is safe to burn trash in the back yard or to run an automobile without smog-control, by law we delegate the details to bureaus. The result is administrative law, which is rightly feared for an ancient reason – *Quis custodiet ipsos custodes?* – "Who shall watch the watchers themselves?" John Adams said that we must have "a government of laws and not men." Bureau administrators, trying to evaluate the morality of acts in the total system, are singularly liable to corruption, producing a government by men, not laws.

Prohibition is easy to legislate (though not necessarily to enforce); but how do we legislate temperance? Experience indicates that it can be accomplished best through the mediation of administrative law. We limit possibilities unnecessarily if we suppose that the sentiment of *Quis custodiet* denies us the use of administrative law. We should rather retain the phrase as a perpetual reminder of fearful dangers we cannot avoid. The great challenge facing us now is to invent the corrective feedbacks that are needed to keep custodians honest. We must find ways to legitimate the needed authority of both the custodians and the corrective feedbacks.

Supplemental Readings for Chapter 3

The Philosophical Basis of Engineering Codes of Ethics

ALBERT FLORES

While the need for engineering codes of conduct might seem readily apparent, they are not so obvious once one starts to mount arguments for their existence. Albert Flores, one of our most renown philosophers of engineering, takes a crack at searching for the philosophical underpinnings of our codes of ethics. Contrast his arguments with those of John Ladd in the following reading. Some of the footnotes have been deleted for the sake of clarity and brevity.

What is the philosophical basis of engineering codes of ethics? The importance of this question is obvious, for without an acceptable answer, we can hardly expect that engineers would willingly submit to the requirements that these codes impose on their conduct. The injunctions that are contained in these codes are intended to restrict engineering practice to areas that are morally acceptable, hence they involve a direct limitation of an engineer's freedom. Obviously, our freedom is one of our most cherished possessions, and its restriction is permissible and acceptable only if there exist compelling reasons that justify such a restriction. Indeed, engineering codes could claim little authority over engineering practice unless there are good reasons to accept their precepts and incorporate them into one's routine activities. At issue, then, is the question: Why should engineers adopt their profession's code of conduct?

The question being asked is not a question about why engineers might be *motivated* to adopt professional codes, nor is it aimed at providing an explanation of the *causes* that account for engineers' acceptance of these codes.

The entire article is in James H. Schaub and Sheila K. Dickinson, eds., *Engineering and Humanities* (Robert E. Krieger, Malabar, FL, 1982). Used with permission.

These are, no doubt, interesting questions, and their answers are clearly relevant to questions concerning why engineers would willingly restrict their professional practice to the limits prescribed in their ethical codes. So, for example, engineers may be motivated to observe these limits because they fear the censure of their colleagues or because claiming allegiance to a code of ethics may enhance their reputation. They may believe that following its injunctions will relieve them of the guilt feelings they have when they fail to obey its prescriptions or that, generally, obedience will bring them rewards, make them happy, or allow them to avoid the responsibilities associated with making their own decisions. These are just some of the explanations that we can construct to account for engineers' behavior, but none can *justify* this behavior. For they all refer to the causes of behavior and not to the reasons why these behaviors are chosen. Questions that ask Why? in reference to how people act, are ambiguous in that they fail to adequately distinguish between reasons and causes. Unfortunately, our conduct is as much governed by motives, feelings, and causes as it is by reasons. And it is with the latter that we are here concerned. In brief, the kind of answer we are seeking is one that justifies adopting a code of ethics, and justifications are constructed from reasons and arguments, not by reference to motives and causes.

In moral philosophy the adoption of a moral code of conduct has been justified in either of two general ways: arguments that refer to the beneficial consequences that flow from following such a code illustrate what is known as a consequentialist justification. John Stuart Mill's extended argument in *On Liberty* showing why it is socially beneficial to respect the freedom of individuals to pursue their own interests unfettered by society, so long as they cause no harm to others, is a paradigm of a consequentialist justification. On the other hand, there are nonconsequentialist, or deontological, arguments (from the Greek "deon" meaning "ought"). These arguments appeal to principles of obligation or rules expressing duties in order to justify adopting a particular code. Accordingly, Kant has argued that no set of beneficial consequences can, for example, justify the telling of a lie, because lying violates a basic moral obligation or duty that we have to be truthful in word and deed. It is, in short, in the nature of the act of lying itself and not in its possible harmful consequences that we find justification for its prohibition, according to those who adopt this deontological mode of justification. It would, however, be mistaken to conclude that truth-telling is not also justified by an appeal to the good consequences that following such a principle would produce. Indeed, many of our most cherished moral principles are justified in either of these two ways. In attempting to provide an answer to the question of why engineers should adopt their professional codes of conduct, we will examine arguments illustrating both these lines of justification.

However, before we proceed to this task, there is an important preliminary task to which we must now attend. Although engineering codes of ethics purport to define the standards of ethical performance expected of professional engineers, in reality these codes contain numerous elements which have nothing to do with ethics. In addition to stating the moral duties which engineers are expected to honor, these codes contain principles addressing business practices, professional etiquette, and the special interests of the profession that are intended to regulate the behavior of engineers, but not in areas that can be reasonably called areas of *ethical* concern. Questions concerning whether an engineer should or should not advertise or engage in competitive bidding, for example, do not inherently raise moral issues, though they are certainly concerns which the engineering profession has an interest in overseeing. Whether it is wise to include such nonmoral concerns in a code of "ethics" is an engaging, but separate, matter. In any case, since our interest here is in understanding the reasons why engineers should adopt the *moral* principles embodied in these codes, it is important that we distinguish moral constraints from those nonmoral injunctions the codes contain. Thus the justificatory arguments which we shall discuss are aimed only at supporting those sections of the profession's codes that deal with matters of ethics.

This objective is complicated by the fact that in contrast to most major professions, engineering is organized into over one hundred and fifty separate societies and professional associations, and nearly all these organizations have enacted separate codes of ethics for the members. Hence, there is no single code of ethics that is "the profession's" code. But although the existence of a number of distinct codes can be a source of some important difficulties for a profession, these codes are fortunately similar in the respect that they all incorporate a set of basic ethical principles that are easily identifiable. They include the following negative and positive duties. Engineers are obliged to use their skills and knowledge in ways that do not injure, harm, or endanger the welfare and safety of the public, their clients, or employers. They are obliged, in addition, to be honest, fair, and loyal in all their dealings. And finally, they are expected to practice their profession with the dignity and honor befitting their special status as professionals, respecting the moral integrity of all those with whom they deal. In short, engineers are morally required to do no harm and to pursue a set of positive moral values in the routine conduct of their professional practice. These summarize the moral duties that engineering codes contain.

According to the deontological approach to ethics, the moral principles that are contained in engineering codes are justified only if they conform with our *duty* as moral agents. Kant, for example, argued that obligation or duty is the basic moral concept, and it is in terms of this concept that the moral rightness or wrongness of an action is defined. An action is "right" provided that it conforms with our moral duty and "wrong" when it is contrary to duty. Implicit in each is the presumption that we should do what is right and avoid doing wrong. And this is just another way of expressing what has come to be known as taking "the moral point of view." Since our concern is with justifying the ethical principles in engineering codes, can it be shown that the principles mentioned . . . *should* be followed because they direct us to do what is morally right? That is, do they express what are our duties as prescribed by a rationally conceived morality? If we can show this, we would then have a sufficient reason to justify adopting these codes.

To begin with, consider the negative duty to do no harm in the application of engineering skills. Although this duty is normally stated in positive terms, as a duty to protect the public's safety, health, and welfare, it clearly intends that engineers should avoid using their skills in ways that may harmfully jeopardize the public's safety, health, or welfare. Obviously, we all believe and feel that it is wrong to cause harm to another, but what reasons ground this belief?

An important feature of Kant's moral theory is that the rightness or wrongness of an act is an intrinsic property of the act itself, and hence its moral quality is independent of its effectiveness to bring about desired ends or the special interests of those who perform it. Moreover, morality, as conceived by Kant, is rational and universal. That is, any action which is right or wrong at all, would be right or wrong for any rational individual, regardless of the circumstances. Hence if it is wrong for engineers to cause harm to another, then it is wrong for anyone to cause harm to another. And the wrongness of such an action rests precisely in the fact that we cannot consistently will that everyone can freely do harm to another and at the same time continue to pursue causing such harm ourselves. In other words, a morally acceptable principle of conduct is one that is characterized by the logical rule that if anyone accepts it as a principle justifying their action, then rational consistency demands that it must be acceptable that *everyone* else make it their principle of conduct, too. This rule, which Kant calls the *categorical imperative,* is put forth as the basic criterion for judging the morality of reasons or principles. Thus it requires that any reason or principle offered to justify an action must be capable of being "universalized"; that is, it must be capable of being willed as a universal law of morality. And if it fails the test, then morality as rationally conceived imposes upon us a duty to avoid such an action, because it is morally wrong. Clearly we cannot will that it is acceptable that everyone is free to cause harm to another, without inconsistency, since no one can rationally accept as a moral principle that others are morally justified in harming them. So, it is everyone's moral duty to do no harm, and by implication this is a duty which engineers are required by ethics to accept, as well.

A similar argument can be constructed for each of the positive duties that engineering codes of ethics require of engineers. Thus, for example, engineers are enjoined to be honest in their relations with clients, colleagues, employers, and with the public at large. Honesty simply means being truthful in word and deed, or avoiding falsehood and deceit when we speak and act. Does the duty to be honest satisfy the categorical imperative; that is, can we will it as a universal principle of morality that everyone be honest, without inconsistency? To see that we can, it is helpful to examine its denial, for if allowing everyone to be *dis*honest violates the demands of rational consistency, then dishonesty is wrong, and, conversely, honesty must be morally right, hence our duty. The categorical imperative implies that *if* I am justified in being deceitful in my relations with others, then everyone else is also justified to act on this ground. But such a universal principle would make communication impossible, either by word or deed, for no one could know whether what is said or done is true or not. And if there can be no communication, then we cannot in principle be deceitful, since deception presupposes the possibility of successful communication in order to convey to another what is untrue. But, of course, the possibility of communication is exactly what is denied by implication when dishonesty is universalized. Thus dishonesty is morally wrong, and it is, therefore, our duty to avoid being dishonest. In brief, morality requires each of us to be honest in our relations with others.

And if this is what is required of every rational individual, then it is likewise required of professional engineers. It is, therefore, appropriate that engineering codes impose a moral obligation of honesty on engineers and reasonable that engineers adopt such a code. With some effort we can similarly establish the moral validity of those other positive duties to be loyal and fair which the profession's codes contain.

It should be clear by now that one of the distinctive marks of this deontological, or nonconsequentialist, approach to the justification of codes of conduct is its logical formalism. In other words, an important consideration which this approach emphasizes is that acceptable moral principles must meet the test of rational consistency. However, it is not just consistency with one's self, but of necessity it must encompass the entire community of freely choosing moral agents if it is to be a code which is rationally acceptable and morally valid.

But the injunction that engineers should "uphold and advance the integrity, honor and dignity of the engineering profession" appears to be merely a provincial consideration and not an injunction that we can reasonably expect *every* member of the moral community to observe. Yet it would be strange to deny that engineers have these kinds of moral obligations to their profession. Can we justify these duties in a manner similar to the arguments offered above?

Not strictly, although a nonconsequentialist justification of "institutional objectives" can be constructed in terms of the role-defined duties required of individuals who are members of institutions, such as the engineering profession, provided that the institution itself can be shown to contribute to the well-being of society. So, for example, although the value of the engineering profession as an institution can be established by reference to the value of the goods it produces for society, the special duties that engineers have to their profession are justified not by their functional utility but arise as a logical consequence of the fact that these individuals are members of a socially important institution that imposes on its members obligations critical to fulfilling its purpose or function. Like most institutions, the engineering profession is constituted by rules that define the activities which its members are required to perform in order to satisfactorily accomplish its social function or purpose. By accepting membership into a profession or any other institution, one voluntarily takes on a set of role-defined duties above and beyond one's general moral obligations, which nonmembers are not expected to satisfy. So, engineers' duties to uphold the honor and dignity of their profession may be viewed as duties that are logically required by the very concept of a "professional engineer." Thus it is in virtue of one's role as a professional that we find a morally acceptable reason for the professional duties engineering ethical codes impose on the their members.

Except for the argument just given, the arguments offered above are formally representations of the Golden Rule which requires that "we do unto others as we would have them do unto us." Although this is clearly a persuasive reason for incorporating the profession's ethical codes into one's practice, we can as well appeal to the good consequences that flow from these principles of conduct in order to further substantiate accepting their injunctions.

And what could be more plausible than that our moral duty requires of us to promote the greatest balance of good over evil? That is, the rightness of an action is justified by reference to the good consequences it produces, and an action that on the whole produces bad consequences is, consequently, morally wrong. Whereas according to the deontologist an action is right or wrong insofar as it conforms with our moral duty, according to this consequentialist approach, the rightness or wrongness of actions is a function of the goodness or badness of its consequences. Mill has, for example, defended the following principle as a way of determining the moral rightness of actions: "actions are right in proportion as they tend to promote happiness, wrong as they tend to produce the reverse of happiness." Although Mill uses the term "happiness" to indicate pleasure and the absence of pain, it is, perhaps, more appropriate to follow Aristotle's notion of happiness as the well-being or good of the individual, or society in general. Thus, according to this "principle of utility," the duty of every moral agent is the promotion of the greatest good for the greatest number. At issue, then, is whether the moral injunctions contained in the engineering profession's codes of ethics contribute to this end.

Although it is admittedly difficult to calculate all the various consequences that result from engineering activities, it is reasonable to assert that if these activities are characterized by honesty, fairness, and loyalty, far greater good would result than if they were not. Indeed, not only does good result from being honest, etc., but much evil is avoided, as well. Furthermore, there is no doubt that pursuing the practice of engineering so that one avoids causing injury, harm, or death to others is morally justified by the fact that such harm is an inherent evil, while intentionally avoiding it is inherently productive of much good. To put the matter more generally, engineers should follow the morally relevant injunctions contained in their profession's codes of ethics because the effect of such a conduct is a substantial increase in the general well-being of society and it avoids countless bad consequences that would naturally result if such principled conduct were not followed or consistently rejected. Thus the philosophical basis of engineering codes of ethics is found not only in their consistency with the duties morality requires of us all but also in the goodness of the consequences that result from restricting engineering practice to the limits they set down.

Although the philosophical ground of engineers' codes is rather straightforward, this should not mislead us to conclude that so long as we strictly abide by its provisions our actions will be rationally or morally justified. Nor should we conclude that we will be able to know what we should and should not do in specific cases. Engineers may have moral obligations to protect the public's welfare, to be honest and loyal, etc., but these are only *prima facie* duties. That is, they describe engineers' actual duties if and only if all other things are equal such that there are no special qualifying, morally relevant circumstances that would justify overriding them, in a particular case. To see what specifically is meant here, consider the following example:

> A salaried production engineer discovers in the routine performance of his duties that an inferior grade of material is being used in the manufacture of a product he supervises. Engineering experience tells him that the safety of the product will be seriously compromised unless it is produced with the quality of material identified in its specifications. However, after bringing this to the attention of management, he learns that it is precisely the economic benefit that derives from using this lower-grade material that allows his employers to remain competitive in the marketplace. In short, it is highly unlikely that they will change the grade of the material to meet the product's specifications, because to do so would have very serious negative economic effects on the manufacturer's profitability. Management appears, moreover, willing to accept the risks of product liability suits because it is more cost efficient than using the required materials.

What should the engineer in this case do?

If we look to the profession's codes of ethics, the answer may seem evident: the engineer has a duty to protect the public's health and welfare. But, in addition, he is obliged to be loyal to colleagues and employer; to practice his profession with dignity and integrity; and to be competent in the performance of his duties. Clearly, in this case these duties come into direct conflict with each other such that the engineer is faced with a moral dilemma involving conflicting obligations to the public, his employer, and his profession, which the profession's codes only intensify, but do not answer. Since the engineer cannot simultaneously satisfy all the demands that his profession's codes make on him in this situation, the interesting question it poses is how should this dilemma be resolved so that we know what our actual duty is in cases like this.

On this question, however, the codes are silent. And that is as it should be, because insofar as they are written in general terms to provide a common guide for professional practice, professional codes cannot be expected to function as an algorithm that gives specific answers to every practical moral problem an engineer might confront. But the lesson this suggests is that not only are there good reasons for engineers to accept the moral requirements dictated by their profession's codes of ethics, but it is also important that engineers understand the nature of the philosophical ground justifying these codes if they are to know how to deal effectively with the kinds of ethical problems posed by this and similar cases. Without this understanding, engineers are likely to continue to perceive their professional codes as just so many more rules that limit and infringe on their freedom, presenting ethical problems that have, unfortunately, been easier to ignore than to confront.

The Quest for a Code
of Professional Ethics:
An Intellectual and Moral Confusion

JOHN LADD

Ladd begins this paper by asserting that his role is to act as a gadfly – which is a legitimate function of philosophers. Gadflies are
often wrong . . . but never in doubt! So it is with Ladd.

What he is proposing is that the entire concept of codes of
ethics is nonsense. He demonstrates his ideas using compelling
arguments, and one is almost sucked into the trap. Recognize,
however, that codes of ethics were invented by the professionals
not only to establish their own rules of conduct with regard to
each other, but also to push the decision making toward the public good. If we accept Ladd's arguments, then we have to ask if
the situation would be improved if we did not have codes of
ethics for engineers and other professionals. If the answer is
"yes," then he has something. If it is "no," then he has little to
offer that would make the world a better place. Judge for yourself.

My role as a philosopher is to act as a gadfly. If this were Athens in the fifth
century B.C. you would probably throw me in prison for what I shall say, and
I would be promptly condemned to death for attacking your idols. But you
can't do that in this day and age; you can't even ask for your money back,
since I am not being paid. All that you can do is to throw eggs at me or simply walk out!

Paper delivered at the AAAS Workshop on Professional Ethics, November 1979, published in *AAAS
Professional Ethics Project*, edited by Rosemary Chalk, Mark S. Frankel, and Sallie B. Chafer (AAAS,
Washington, DC, 1980). Used with permission.

My theme is stated in the title: it is that the whole notion of an organized professional ethics is an absurdity – intellectual and moral. Furthermore, I shall argue that there are few positive benefits to be derived from having a code and the possibility of mischievous side effects of adopting a code is substantial. Unfortunately, in the [space] allotted to me I can only summarize what I have to say on this topic.

(1) To begin with, ethics itself is basically an open-ended, reflective and critical intellectual activity. It is essentially problematic and controversial, both as far as its principles are concerned and in its application. Ethics consists of issues to be examined, explored, discussed, deliberated, and argued. Ethical principles can be established only as a result of deliberation and argumentation. These principles are not the kind of thing that can be settled by fiat, by agreement or by authority. To assume that they can be is to confuse ethics with law-making, rule-making, policy-making and other kinds of decision-making. It follows that, ethical principles, as such, cannot be established by associations, organizations, or by a consensus of their members. To speak of codifying ethics, therefore, makes no more sense than to speak of codifying medicine, anthropology or architecture.

(2) Even if substantial agreement could be reached on ethical principles and they could be set out in a code, the attempt to impose such principles on others in the guise of ethics contradicts the notion of ethics itself, which presumes that persons are autonomous moral agents. In Kant's terms, such an attempt makes ethics heteronomous; it confuses ethics with some kind of externally imposed set of rules such as a code of law, which, indeed, is heteronomous. To put the point in more popular language: ethics must, by its very nature, be self-directed rather than other-directed.

(3) Thus, in attaching disciplinary procedures, methods of adjudication and sanctions, formal and informal, to the principles that one calls "ethical" one automatically converts them into legal rules or some other kind of authoritative rules of conduct such as the bylaws of an organization, regulations promulgated by an official, club rules, rules of etiquette, or other sorts of social standards of conduct. To label such conventions, rules and standards "ethical" simply reflects an intellectual confusion about the status and function of these conventions, rules and standards. Historically, it should be noted that the term "ethical" was introduced merely to indicate that the code of the Royal College of Physicians was not to be construed as a criminal code (i.e., a legal code). Here "ethical" means simply non-legal.

(4) That is not to say that ethics has no relevance for projects involving the creation, certification and enforcement of rules of conduct for members of

certain groups. But logically it has the same kind of relevance that it has for the law. As with law, its role in connection with these projects is to appraise, criticize and perhaps even defend (or condemn) the projects themselves, the rules, regulations and procedures they prescribe, and the social and political goals and institutions they represent. But although ethics can be used to judge or evaluate a disciplinary code, penal code, code of honor or what goes by the name of a "code of ethics," it cannot be identified with any of these, for the reasons that have already been mentioned.

Some General Comments on Professionalism and Ethics

(5) Being a professional does not automatically make a person an expert in ethics, even in the ethics of that person's own particular profession – unless of course we decide to call the 'club rules' of a profession its ethics. The reason for this is that there are no experts in ethics in the sense of expert in which professionals have a special expertise that others do not share. As Plato pointed out long ago in the *Protagoras,* knowledge of virtue is not like the technical knowledge that is possessed by an architect or shipbuilder. In a sense, everyone is, or ought to be, a teacher of virtue; there are no professional qualifications that are necessary for doing ethics.

(6) Moreover, there is no special ethics belonging to professionals. Professionals are not, simply because they are professionals, exempt from the common obligations, duties and responsibilities that are binding on ordinary people. They do not have a special moral status that allows them to do things that no one else can. Doctors have no special right to be rude, to deceive, or to order people around like children, etc. Likewise, lawyers do not have a special right to bend the law to help their clients, to bully witnesses, or to be cruel and brutal – simply because they think that it is in the interests of their client. Professional codes cannot, therefore, confer such rights and immunities; for there is no such thing as professional ethics immunity.

(7) We might ask: do professionals, by virtue of their special professional status, have special duties and obligations over and above those they would have as ordinary people? Before we can answer this question, we must first decide what is meant by the terms "profession" and "professional," which are very loose terms that are used as labels for a variety of different occupational categories. The distinctive element in professionalism is generally

held to be that professionals have undergone advanced, specialized training and that they exercise control over the nature of their job and the services they provide. In addition, the older professions, lawyers, physicians, professors and ministers typically have clients to whom they provide services as individuals. (I use the term "client" generically so as to include patients, students, and parishioners.) When professionals have individual clients, new moral relationships are created that demand special types of trust and loyalty. Thus, in order to answer the question, we need to examine the context under which special duties and obligations of professionals might arise.

(8) In discussing specific ethical issues relating to the professions, it is convenient to divide them into issues of *macro-ethics* and *micro-ethics*. The former comprise what might be called collective or social problems, that is, problems confronting members of a profession as a group in their relation to society; the latter, issues of micro-ethics, are concerned with moral aspects of personal relationships between individual professionals and other individuals who are their clients, their colleagues and their employers. Clearly the particulars in both kinds of ethics vary considerably from one profession to another. I shall make only two general comments.

(9) Micro-ethical issues concern the personal relationships between individuals. Many of these issues simply involve the application of ordinary notions of honesty, decency, civility, humanity, considerateness, respect and responsibility. Therefore, it should not be necessary to devise a special code to tell professionals that they ought to refrain from cheating and lying, or to make them treat their clients (and patients) with respect, or to tell them that they ought to ask for informed consent for invasive actions. It is a common mistake to assume that *all* the extra-legal norms and conventions governing professional relationships have a moral status, for every profession has norms and conventions that have as little to do with morality as the ceremonial dress and titles that are customarily associated with the older professions.

(10) The macro-ethical problems in professionalism are more problematic and controversial. What are the social responsibilities of professionals as a group? What can and should they do to influence social policy? Here, I submit, the issue is not one of professional roles, but of *professional power*. For professionals as a group have a great deal of power; and power begets responsibility. Physicians as a group can, for instance, exercise a great deal of influence on the quality and cost of health care; and lawyers can have a great deal of influence on how the law is made and administered, etc.

(11) So called "codes of professional ethics" have nothing to contribute

either to micro-ethics or to macro-ethics as just outlined. It should also be obvious that they do not fit under either of these two categories. Any association, including a professional association, can, of course, adopt a code of conduct for its members and lay down disciplinary procedures and sanctions to enforce conformity with its rules. But to call such a disciplinary code a code of *ethics* is at once pretentious and sanctimonious. Even worse, it is to make a false and misleading claim, namely, that the profession in question has the authority or special competence to create an ethics, that it is able authoritatively to set forth what the principles of ethics are, and that it has its own brand of ethics that it can impose on its members and on society.

I have briefly stated the case against taking a code of professional ethics to be a serious ethical enterprise. It might be objected, however, that I have neglected to recognize some of the benefits that come from having professional codes of ethics. In order to discuss these possible benefits, I shall first examine what some of the objectives of codes of ethics might be, then I shall consider some possible benefits of having a code, and finally, I shall point out some of the mischievous aspects of codes.

Objectives of Codes of Professional "Ethics"

In order to be crystal clear about the purposes and objectives of a code, we must begin by asking: to whom is the code addressed? Although ostensibly codes of ethics are addressed to the members of the profession, their true purposes and objectives are sometimes easier to ascertain if we recognize that codes are in fact often directed at other addressees than members. Accordingly, the real addressees might be any of the following: (a) members of the profession, (b) clients or buyers of the professional services, (c) other agents dealing with professionals, such as government or private institutions like universities or hospitals, or (d) the public at large. With this in mind, let us examine some possible objectives.

First, the objective of a professional code might be "inspirational," that is, it might be used to inspire members to be more "ethical" in their conduct. The assumption on which this objective is premised is that professionals are somehow likely to be amoral or submoral, perhaps, as the result of becoming professionals, and so it is necessary to exhort them to be moral, e.g. to be honest. I suppose there is nothing objectionable to having a code for this reason; it would be something like the Boy Scout's Code of Honor, something to frame and hang in one's office. I have severe reservations, however, about whether a code is really needed for this purpose and whether it will do any good; for those to whom it is addressed and who need it the most will not adhere to it anyway, and the rest of the good people in the profession will not need it because they already know what they ought to do. For this reason, many respectable members of a profession regard its code as a joke and as something not to be taken seriously. (Incidentally, for much the same kind of reasons as those just given, there are no professional codes in the academic or clerical professions.)

A second objective might be to alert professionals to the moral aspects of their work that they might have overlooked. In jargon, it might serve to sensitize them or to raise their consciousness. This, of course, is a worthy goal – it is the goal of moral education. Morality, after all, is not just a matter of doing or not doing, but also a matter of feeling and thinking. But, here again, it is doubtful that it is possible [to] make people have the right feelings or think rightly through enacting a code. A code is hardly the best means for teaching morality.

Thirdly, a code might, as it was traditionally, be a disciplinary code or a "penal" code used to enforce certain rules of the profession on its members in order to defend the integrity of the professional and to protect its professional standards. This kind of function is often referred to as "self-policing." It is unlikely, however, that the kind of disciplining that is in question here could be handled in a code of ethics, a code that would set forth in detail criteria for determining malpractice. On the contrary, the "ethical" code of a profession is usually used to discipline its members for other sorts of 'unethical conduct,' such as stealing a client away from a colleague, for making disparaging remarks about a colleague in public, or for departing from some other sort of norm of the profession. (In the original code of the Royal College of Physicians, members who failed to attend the funeral of a colleague were subject to a fine!) It is clear that when we talk of a disciplinary code, as distinguished from an exhortatory code, a lot of new questions arise that cannot be treated here; for a disciplinary code is quasi-legal in nature, it involves adjudicative organs and processes, and it is usually connected with complicated issues relating to such things as licensing.

A fourth objective of a code might be to offer advice in cases of moral perplexity about what to do: e.g. should one report a colleague for malfeasance? Should one let a severely defective newborn die? If such cases present genuine perplexities, then they cannot and should not be solved by reference to a code. To try to solve them through a code is like trying to do surgery with a carving knife! If it is not a genuine perplexity, then the code would be unnecessary.

A fifth objective of a professional code of ethics is to alert prospective clients and employers to what they may and may not expect by way of service from a member of the profession concerned. The official code of an association, say, of engineers, provides as authoritative statement of what is proper and what is improper conduct of the professional. Thus, a code serves to protect a professional from improper demands on the part of employer or client, e.g. that he lie about or coverup defective work that constitutes a public hazard. Codes may thus serve to protect 'whistle-blowers.' (The real addressee in this case is the employer or client.)

Secondary Objectives of Codes – Not Always Salutary

I now come to what I shall call "secondary objectives," that is, objectives that one might hesitate always to call "ethical," especially since they often provide an opportunity for abuse.

The first secondary objective is to enhance the image of the profession in the public eye. The code is supposed to communicate to the general public (the addressee) the idea that the members of the profession concerned are service oriented and that the interests of the client are always given first place over the interests of the professional himself. Because they have a code they may be expected to be trustworthy.

Another secondary objective of a code is to protect the monopoly of the profession in question. Historically, this appears to have been the principal objective of a so-called code of ethics, e.g. Percival's code of medical ethics. Its aim is to exclude from practice those who are outside the professional in-group and to regulate the conduct of the members of the profession so as to protect it from encroachment from outside. Sometimes this kind of professional monopoly is in the public interest and often it is not.

Another secondary objective of professional codes of ethics, mentioned in some of the literature, is that having a code serves as a status symbol; one of the credentials for an occupation to be considered a profession is that it have a code of ethics. If you want to make your occupation a profession, then you must frame a code of ethics for it: so there are codes for real estate agents, insurance agents, used car dealers, electricians, barbers, etc., and these codes serve, at least in the eyes of some, to raise their members to the social status of lawyers and doctors.

Mischievous Side-Effects of Codes of Ethics

I now want to call attention to some of the mischievous side-effects of adopting a code of ethics:

The first and most obvious bit of mischief, it that having a code will give a sense of complacency to professionals about their conduct. "We have a code of ethics," they will say, "so everything we do is ethical." Inasmuch as a code, of necessity, prescribes what is minimal, a professional may be encouraged by the code to deliver what is minimal rather than the best that he can do. "I did everything that the code requires . . . "

Even more mischievous than complacency and the consequent self-congratulation, is the fact that a code of ethics can be used as a cover-up for what might be called basically "unethical" or "irresponsible" conduct.

Perhaps the most mischievous side-effect of codes of ethics is that they tend to divert attention from the macro-ethical problems of a profession to its micro-ethical problems. There is a lot of talk about whistle-blowing. But it concerns individuals almost exclusively. What is really needed is a thorough scrutiny of professions as collective bodies, of their role in society and their effect on the public interest. What role should the professions play in determining the use of technology, its development and expansion, and the distribution of the costs (e.g. disposition of toxic wastes) as well as the benefits of technology? What is the significance of professionalism from the moral point of view for democracy, social equality, liberty and justice? There are lots of ethical problems to be dealt with. To concentrate on codes of ethics as if they represented the real ethical problems connected with professionalism is to capitulate to *struthianism* (from the Greek word *struthos* = ostrich).

One final objection to codes that needs to be mentioned is that they inevitably represent what John Stuart Mill called the "tyranny of the majority" or, if not that, the "tyranny of the establishment." They serve to and are designed to discourage if not suppress the dissenter, the innovator, the critic.

By way of conclusion, . . . a few words about what an association of professionals can do about ethics. On theoretical grounds, I have argued that it cannot codify an ethics and it cannot authoritatively establish ethical principles or prescribed guidelines for the conduct of its members – as if it were *creating* an ethics! But there is still much that associations can do to promote further understanding of and sensitivity to ethical issues connected with professional activities. For example, they can fill a very useful educational function by encouraging their members to participate in extended discussions of issues of both micro-ethics and macro-ethics, e.g. questions about responsibility; for these issues obviously need to be examined and discussed much more extensively than they are at present – especially by those who are in a position to do something about them.

What Are Codes of Ethics For?

JUDITH LICHTENBERG

John Ladd is all wet – or so asserts Judith Lichtenberg, a philosopher at the University of Maryland. In fact, codes of ethics create the descriptions of what proper behavior ought to be, and even though one may not agree with all provisions of a code of ethics, if that code is well constructed, it can serve as a guide to proper behavior. With time, asserts Lichtenberg, beliefs will follow behavior, and a better society results. Perhaps this is too much to ask of a code of ethics, but there seem to be more reasons for having them than for not incorporating codes of ethics into our professions.

John Ladd has argued that the imposition of principles on other people "in guise of ethics contradicts the notion of ethics itself, which presumes that persons are autonomous moral agents." A code of ethics, Ladd believes, by its nature converts ethical issues into something else – matters of legal or other authoritative rules, perhaps, but certainly not ethics. Ethics cannot be imposed from without.

Whatever appeal Ladd's view possesses derives from an exaggerated emphasis on the word "ethics." If one insists that to act ethically is necessarily or by definition to act autonomously, and not therefore to obey rules externally imposed, then a code of ethics is an oxymoron, precluded from the start. Ladd apparently believes that the mere articulation of a code of ethics, irrespective of the attachment of sanctions, is incompatible with ethics in its true meaning.

For some purposes, it is clear, the identification of the ethical with

Excerpted from Judith Lichtenberg, "What Are Codes of Ethics For?," in Margaret Coady and Sidney Block, eds., *Codes of Ethics and the Professions* (Melbourne University Press, Melbourne, 1996). Used with permission.

autonomous, freely chosen action is appropriate. Ethics is centrally con-
sented with the cultivation of moral character. We care about people's
motives, desires, wills. We want people to do the right things for the right
reasons. We may care about character intrinsically, because, say, as Kantians
or Christians we think that a good will or a pure heart is supremely important. We may also care about it instrumentally, because we think that the
more efficient way to reproduce good outcomes is to produce agents with
good dispositions. I believe that we care about character, and that it is right
to care about it, for both kinds of reasons: a human being with certain kinds
of motives and desires is intrinsically valuable, and that sort of person tends
to produce better consequences in the world.

Let us leave aside for the moment the question of how a code of ethics
could play some role in influencing people's dispositions or characters. It
seems clear that one primary purpose of a code is simply to increase the
probability that people will behave in some ways rather than others. If this
is so, then a code of ethics may be both possible and effective, just as a system of law is. A code of ethics can give people a reason – perhaps a decisive
reason – to act in one way rather than another. Their motives might lack
moral purity; they might comply with the code for fear of the sanctions of
disobedience. That would be regrettable, in the sense that we would prefer
people to act virtuously for virtue's sake. But it doesn't follow that a code of
ethics is either impossible or undesirable. If what Ladd objects to is simply
the word "ethics," nothing significant is lost by speaking of a code of conduct instead.

So, in other words, sometimes we care about people's autonomy and
sometimes we don't; sometimes we care more and sometimes less. It
infringes autonomy to prohibit and punish murder, but that does not figure
as a plausible objection to laws against murder. The question comes down to
the relative importance to us – to society – of outward behavior compared to
the reason or motive for it or the character of the agent. Insofar as we are
concerned to produce the behaviour, we care less about people's autonomy;
insofar as we care about people's reasons or their character, their freely chosen decision to act – their autonomy – is essential. This is not the matter of
either/or; we can desire both. Yet there may be an inherent conflict in
achieving both simultaneously, or at least in knowing whether both are
achieved, because external pressures to comply render the autonomy of the
act more doubtful. My own view is that for most purposes – although by no
means all – we are interested primarily in securing outward compliance.

There are, of course, many reasons why it is better for people not simply to act rightly, but to do so for the right reasons – not the least of which is that (to put it crudely) self-regulation is generally more cost-effective than reliance on external sanctions. Train people's habits and dispositions and you don't need to monitor their behavior so closely. But this is not an argument against codes of ethics, unless it is also an argument against law. If the anarchist objection to the obligation to obey the law can be answered, so can the analogous objection to codes of ethics.

Ladd's view might be summed up in the slogan "You ought not legislate morality." This is different from the usual understanding of "You can't legislate morality," which means not that doing so is undesirable but that it is ineffective. The point is related to Ladd's, however, for the idea is that the moral life is an inner life not accessible to the manipulation of outward behaviour. As a claim about how people's attitudes develop and change, however, this is simply false. Examples such as the civil rights movement demonstrate that changes in the law can over time significantly change people's attitudes; one era's conventional wisdom is an embarrassment to the next. Even if our ultimate aim is to change people's characters, desires and reasons for acting, then, we could do a lot worse than begin by manipulating their incentives to act by requiring certain behaviour and attaching penalties to non-compliance.

In Ladd's view, the truly ethical person could in the nature of things pay no heed to a code of ethics, because an autonomous being cannot be guided by externally given rules. A different objection to codes of ethics is that they are either unnecessary or useless. They are unnecessary because good people don't need them. Good people know how to act and are motivated accordingly. They need codes neither for instructional purposes nor as external incentives. Bad people will not be moved to comply with codes, except by implausibly harsh and certain sanctions. So codes are either unnecessary or useless.

This view makes at least two implausible assumptions. One is that determining what is right is always easy for a person of good character. The other is that the world divides neatly into the virtuous and the vicious, or that it divides neatly into the virtuous and the non-virtuous. Most people fall short of our ideal of a good person (it is rare not to be disappointed morally in many if not most of the people some of the time), but they fall short in varying degrees. And most can be moved by a variety of methods short of formal sanctions. Fear of the disapproval of our peers is the most obvious.

It is not necessary to assume that evil is rampant or that human nature is nasty and brutish to see that many people who would not do anything

grossly immoral easily stray from the virtuous path under certain conditions. When the temptations are significant, when the price of adherence (in
terms, for example, of the sacrifice to our interest) is high, when the social
consequences of violation (harm to others) are relatively slight, when the
costs of violation are low – under such circumstances it is easy to be led from
doing just what you ought do (assuming that what you ought to do is clear).
With or without sanctions, a code of ethics can give people "within the normal ethical range," as we might say, a reason to do what they might not be
sufficiently moved to do on their own. Let me suggest several ways it might
do this.

The first way causes people to redescribe the nature of the situation confronting them. People may do wrong for reasons other than weakness of will
or ignorance. The problem may be that they have not brought to explicit
consciousness the character of what they are doing or not doing. It would be
a mistake to say they don't *know* that what they are doing is wrong, for the
problem is not one of simple ignorance. It is rather that they have not
thought about just what it is they are doing; they have not described it to
themselves properly, if at all.

Sexual behaviour, with its nearly limitless potential for self-deception and
other forms of psychological cover, provides the most obvious source of
examples. Professors who regularly make passes at their students may not
see what they are doing as a violation of their professional responsibility. It
is not that they have formulated a description of what they are doing and
actively justify it. Rather, they have not had the occasion – and have not had
the occasion forced upon them – to describe to themselves what they are
doing. Were they to do so they would probably feel some discomfort, for the
description itself is almost inevitably laden with moral overtones. It would
be odd to describe such people as suffering from ignorance – "They don't
know that getting into these relationships is wrong." It's more plausible to
say "They've never thought about it, or never thought about it like that; if
they did they would see a problem." This is not ignorance in the usual sense,
for ignorance would survive the coming of consciousness; if I am ignorant
that what I am doing is wrong, then even if I describe what I am doing to
myself, I feel no discomfort. What we have here is not ignorance but a failure to think about what one is doing. A code of ethics can increase the probability that people will think about it – can make it more difficult to engage
in self-deceptive practices – by explicitly describing behaviour that is undesirable or unacceptable.

Two objections to this view might be raised. First, it might seem naive to think that the professor's problem is that he doesn't see what he is doing in the proper way, or that simply by having the behaviour described he would come to see it differently and so change his behaviour. Why wouldn't he instead reject the description offered?

The answer is that the processes at work are subtle and complicated. Changes occur slowly and by degrees, and sometimes they do not occur at all. But people do change not only their behaviour but how they view it. A revolution has occurred in the way we think about sexual harassment, and it has come about partly through this kind of description and redescription. Sexual harassment hasn't disappeared, but it has probably declined, and many men think about the way they interact with women differently from the way they used to. (Much the same could be said about racial interaction.) There are things they would have done or remarks they would have made ten or twenty years ago that they no longer would. This not always simply because they see that others do not find such things acceptable, but because they themselves no longer do either.

One might object to this view – that a code of ethics can make people see what they are doing in a new light – in a slightly different way. These descriptions seem to be morally loaded. Of course, one might say, if you agree to the description of a piece of behaviour then you will agree that it is wrong; the controversy comes in deciding on an accurate description. So a man who engages in behaviour that others think of as sexual harassment and therefore wrong is unlikely to come to an easy agreement with them about how to describe what he is doing. They all agree that sexual harassment is wrong, but they disagree about whether this is an instance of it.

Clearly, some of the controversy comes here. Nevertheless, people come to view their behaviour in a different light partly by encountering new descriptions of it, and such descriptions need not be morally loaded in the way that those who distinguish sharply between factual and evaluative descriptions seem to suggest. Descriptions do not fall neatly into two categories – flat, neutral ones on the one hand and starkly condemnatory or laudatory ones on the other. Our language is much richer than that, allowing for a wide range of subtle difference in tone and value.

By making certain standards of behaviour explicit, then, a code of ethics can make it more difficult to avoid confronting discomforting descriptions of behaviour, and so make it more difficult to continue along certain paths. Codes of ethics do not always do this, of course. For example, the American Medical Association Principles of Medical Ethics used to state that "Sexual misconduct in the practice of medicine violates the trust the patient reposes in the physician and is unethical." Without even a hint of what counts as sexual misconduct, the precept is all but useless, for people can pass over the term comfortably in the belief that what they are doing doesn't count as misconduct. The more recent version of the code leaves less room for interpretation: "Sexual contact that occurs concurrent with the physician–patient relationship constitutes sexual misconduct."

In the cases I have been examining so far, it would be a mistake to describe the problem (for which a code can be a partial solution) as ignorance of the moral truth or of the appropriate moral standards. Rather, people have not thought about what they are doing in a particular way. In other cases, however, it might be accurate to describe the problem as one of ignorance. We don't always know what behaviour is called for in the roles that we choose or that are thrust upon us. Sometimes the fault is ours: we haven't thought sufficiently about what's at stake. Sometimes it's simply that the issue is complicated and defies easy solutions. A code of ethics can embody the accumulated experience and wisdom of many people. To the extent that a code fulfills this function, an answer to one of our initial questions is suggested: a useful code will be detailed and specific. For, from this point of view, we need a code precisely for those situations that are not clear and do not fall out platitudinously.

But this argument for a code might be thought to raise Ladd's objections in pointed form. If resolution of a problem requires extended and deep reflection, isn't the issue sufficiently complex to evoke controversy, and so permit reasonable people to disagree? How, then, can we encode a right answer without infringing individual autonomy?

Consider the alternatives confronting a professional with respect to a detailed code provision. First, she might on reflection come to agree with it. In that case, there is no conflict between the code's dictates and those of conscience. It's worth noting my assumption that those to whom a code applies must reflect on its provisions, rather than comply automatically. The belief that a code of ethics can serve legitimate purposes does not constitute an endorsement of blind obedience. (Ladd's concerns suggest that he mistakes the hold a code has over a person for unthinking compulsion.) A code of ethics, like a legal system, can create a presumption of compliance, but that presumption can be overridden. No external command ever constitutes an absolute and conclusive reason to act.

Second, a person might disagree with the code's provision. That leaves two alternatives: simple disobedience and what we might call conscientious non-compliance, which involves public acknowledgement of disagreement, along with non-compliance. It seems plausible that the more important a code's provision, the stronger the argument that non-compliance should be conscientious. Non-compliance with relatively insignificant provisions raises the same puzzles that non-compliance with lesser laws does. How does one justify speeding? By insisting that one disagrees with the speed limit? And that one is willing to universalize one's disobedience? Or does one say that speeding is a moral failing albeit a small one? However one answers these questions, similar answers can be given in the case of code non-compliance.

The Code of Ethics of the American Society of Civil Engineers

Fundamental Principles

Engineers uphold and advance the integrity, honor and dignity of the engineering profession by:

1. Using their knowledge and skill for the enhancement of human welfare and the environment;
2. being honest and impartial and serving with fidelity the public, their employers and clients;
3. striving to increase the competence and prestige of the engineering profession; and
4. supporting the professional and technical societies of their disciplines.

Fundamental Canons

1. Engineers shall hold paramount the safety, health and welfare of the public and shall strive to comply with the principles of sustainable development in the performance of their professional duties.
2. Engineers shall perform services only in areas of their competence.
3. Engineers shall issue public statements only in an objective and truthful manner.
4. Engineers shall act in professional matters for each employer or client as faithful agents or trustees, and shall avoid conflicts of interest.
5. Engineers shall build their professional reputation on the merit of their services and shall not compete unfairly with others.
6. Engineers shall act in such a manner as to uphold and enhance the honor, integrity, and dignity of the engineering profession.
7. Engineers shall continue their professional development throughout their careers, and shall provide opportunities for the professional development of those engineers under their supervision.

Guidelines to Practice Under the Fundamental Canons of Ethics

CANON 1. Engineers shall hold paramount the safety, health and welfare of the public and shall strive to comply with the principles of sustainable development in the performance of their professional duties.

   a. Engineers shall recognize that the lives, safety, health and welfare of the general public are dependent upon engineering judgments, decisions and practices incorporated into structures, machines, products, processes and devices.
   b. Engineers shall approve or seal only those design documents, reviewed or prepared by them, which are determined to be safe for public health and welfare in conformity with accepted engineering standards.
   c. Engineers whose professional judgment is overruled under circumstances where the safety, health and welfare of the public are endangered, or the principles of sustainable development ignored, shall inform their clients or employers of the possible consequences.
   d. Engineers who have knowledge or reason to believe that another person or firm may be in violation of any of the provisions of Canon 1 shall present such information to the proper authority in writing and shall cooperate with the proper authority in furnishing such further information or assistance as may be required.
   e. Engineers should seek opportunities to be of constructive service in civic affairs and work for the advancement of the safety, health and well-being of their communities, and the protection of the environment through the practice of sustainable development.
   f. Engineers should be committed to improving the environment by adherence to the principles of sustainable development so as to enhance the quality of life of the general public.

CANON 2. Engineers shall perform services only in areas of their competence.

   a. Engineers shall undertake to perform engineering assignments only when qualified by education or experience in the technical field of engineering involved.
   b. Engineers may accept an assignment requiring education or experience outside of their own fields of competence, provided their services are restricted to those phases of the project in which they are

qualified. All other phases of such project shall be performed qualified associates, consultants, or employees.

   c. Engineers shall not affix their signatures or seals to any engineering
plan or document dealing with subject matter in which they lack
competence by virtue of education or experience or to any such
plan or document not reviewed or prepared under their supervisory
control.

CANON 3. Engineers shall issue public statements only in an objective and
truthful manner.

   a. Engineers should endeavor to extend the public knowledge of engineering and sustainable development, and shall not participate in
the dissemination of untrue, unfair or exaggerated statements
regarding engineering.

   b. Engineers shall be objective and truthful in professional reports,
statements, or testimony. They shall include all relevant and pertinent information in such reports, statements, or testimony.

   c. Engineers, when serving as expert witnesses, shall express an engineering opinion only when it is founded upon adequate knowledge
of the facts, upon a background of technical competence, and upon
honest conviction.

   d. Engineers shall issue no statements, criticisms, or arguments on
engineering matters which are inspired or paid for by interested
parties, unless they indicate on whose behalf the statements are
made.

   e. Engineers shall be dignified and modest in explaining their work
and merit, and will avoid any act tending to promote their own
interests at the expense of the integrity, honor and dignity of the
profession.

CANON 4. Engineers shall act in professional matters for each employer or
client as faithful agents or trustees, and shall avoid conflicts of interest.

   a. Engineers shall avoid all known or potential conflicts of interest
with their employers or clients and shall promptly inform their
employers or clients of any business association, interests, or circumstances which could influence their judgment or the quality of
their services.

   b. Engineers shall not accept compensation from more than one party
for services on the same project, or for services pertaining to the

same project, unless the circumstances are fully disclosed to and agreed to, by all interested parties.

c. Engineers shall not solicit or accept gratuities, directly or indirectly, from contractors, their agents, or other parties dealing with their clients or employers in connection with work for which they are responsible.

d. Engineers in public service as members, advisors, or employees of a governmental body or department shall not participate in considerations or actions with respect to services solicited or provided by them or their organization in private or public engineering practice.

e. Engineers shall advise their employers or clients when, as a result of their studies, they believe a project will not be successful.

f. Engineers shall not use confidential information coming to them in course of their assignments as a means of making personal profit if such action is adverse to the interests of their clients, employers or the public.

g. Engineers shall not accept professional employment outside of their regular work or interest without the knowledge of their employers.

CANON 5. Engineers shall build their professional reputation on the merit of their services and shall not compete unfairly with others.

a. Engineers shall avoid not give, solicit or receive either directly or indirectly, any political contribution, gratuity, or unlawful consideration in order to secure work, exclusive of securing salaried positions through employment agencies.

b. Engineers should negotiate contracts for professional services fairly and on the basis of demonstrated competence and qualifications for the type of professional service required.

c. Engineers may request, propose or accept professional commissions on a contingent basis only under circumstances in which their professional judgments would not be compromised.

d. Engineers shall not falsify or permit misrepresentation of their academic or professional qualifications or experience.

e. Engineers shall give proper credit for engineering work to those to whom credit is due, and shall recognize the proprietary interests of others. Whenever possible, they shall name the person or persons who may be responsible for designs, inventions, writings or other accomplishments.

f. Engineers may advertise professional services in a way that does not contain misleading language or is in any other manner derogatory to the dignity of the profession. Examples of permissible advertising are as follows:

Professional cards in recognized, dignified publications, and listings in rosters or directories published by responsible organizations, provided that the cards or listings are consistent in size and content and are in a section of the publication regularly devoted to such professional cards.

Brochures which factually describe experience, facilities, personnel and capacity to render service, providing they are not misleading with respect to the engineer's participation in projects described.

Display advertising in recognized dignified business and professional publications, providing it is factual and is not misleading with respect to the engineer's extent of participation in projects described.

A statement of the engineers' names or the name of the firm and statement of the type of service posted on projects for which they render services.

Preparation or authorization of descriptive articles for the lay or technical press, which are factual and dignified. Such articles shall not imply anything more than direct participation in the project described.

Permission by engineers for their names to be used in commercial advertisements, such as may be published by contractors, material suppliers, etc., only by means of a modest, dignified notation acknowledging the engineers' participation in the project described. Such permission shall not include public endorsement of proprietary products.

g. Engineers shall not maliciously or falsely, directly or indirectly, injure the professional reputation, prospects, practice or employment of another engineer or indiscriminately criticize another's work.

h. Engineers shall not use equipment, supplies, laboratory or office facilities of their employers to carry on outside private practice without the consent of their employers.

CANON 6. Engineers shall act in such a manner as to uphold and enhance the honor, integrity, and dignity of the engineering profession.

a. Engineers shall not knowingly act in a manner which will be derogatory to the honor, integrity, or dignity of the engineering profession or knowingly engage in business or professional practices of a fraudulent, dishonest or unethical nature.

CANON 7. Engineers shall continue their professional development throughout their careers, and shall provide opportunities for the professional development of those engineers under their supervision.

a. Engineers should keep current in their specialty fields by engaging in professional practice, participating in continuing education courses, reading in the technical literature, and attending professional meetings and seminars.
b. Engineers should encourage their engineering employees to become registered at the earliest possible date.
c. Engineers should encourage engineering employees to attend and present papers at professional and technical society meetings.
d. Engineers shall uphold the principle of mutually satisfying relationships between employers and employees with respect to terms of employment including professional grade descriptions, salary ranges, and fringe benefits.

Supplemental Readings for Chapter 4

Justification for the Use of Animals

ST. THOMAS AQUINAS

St. Thomas Aquinas (1225–1274) was both a philosopher and theologian, and was responsible in great part for bringing logic and reason into theological arguments. He set up a hierarchy of intellect, with God at the top as the bearer of pure intellect, followed by humans, and finally by animals. He believed that since animals could not reason, they were significantly inferior to humans and therefore humans could do what they wanted with the animals. He believed that cruelty to animals is wrong, but only because of the effect this has on humans. This argument has been made by other philosophers, including John Locke. An interesting exercise for what follows would be to diagram the arguments used by St. Thomas Aquinas, to see what the premises are and how the conclusions follow from these premises. Keep in mind that he was writing this in the Middle Ages, when certain assumptions about God were accepted as fact.

In the first place then, the very condition of the rational creature, in that it has dominion over its actions, requires that the care of providence should be bestowed on it for its own sake: whereas the condition of other things that have not dominion over their actions shows that they are cared for, not for their own sake, but as being directed to other things. Because that which acts only when moved by another, is like an instrument; whereas that which acts by itself, is like a principal agent. Now an instrument is required, not for its own sake, but that the principal agent may use it. Hence whatever is done for the care of the instruments must be referred to the principal agent as its end, whereas any such action directed to the principal agent as such, either by the

St. Thomas Aquinas, "Difference Between Rational and Other Creatures," in *Summa Contra Gentiles* (Benzinger Brothers, London, 1928). Used with permission.

agent itself or by another, is for the sake of the same principal agent. Accordingly intellectual creatures are ruled by God, as though He cared for them for their own sake, while other creatures are ruled as being directed to rational creatures.

Again. That which has dominion over its own act is free in its action, because he is free who is cause of himself: whereas that which by some kind of necessity is moved by another to act, is subject to slavery. Therefore every other creature is naturally under slavery; the intellectual nature alone is free. Now, in every government provision is made for the free for their own sake; but for slaves that they may be useful to the free. Accordingly divine providence makes provision for the intellectual creature for its own sake, but for other creatures for the sake of the intellectual creature.

Moreover. Whenever certain things are directed to a certain end, if any of them are unable of themselves to attain to the end, they must needs be directed to the end for their own sake. Thus the end of the army is victory, which the soldiers obtain by their own action in fighting, and they alone in the army are required for their own sake; whereas all others, to whom other duties are assigned, such as the care of horses, the preparing of arms, are requisite for the sake of the soldiers of the army. Now, it is clear from what has been said, that God is the last end of the universe, whom the intellectual nature alone obtains in Himself, namely by knowing and loving Him, as was proved above. Therefore the intellectual nature alone is requisite for its own sake in the universe, and all others for its sake.

Further. In every whole, the principal parts are requisite on their own account for the completion of the whole, while others are acquired for the preservation or betterment of the former. Now, of all the parts of the universe, intellectual creatures hold the highest place, because they approach nearest to the divine likeness. Therefore divine providence provides for the intellectual nature for its own sake, and for all others for its sake.

Besides. It is clear that all the parts are directed to the perfection of the whole; since the whole is not on account of the parts, but the parts on account of the whole. Now, intellectual natures are more akin to the whole than other natures; because, in a sense, the intellectual substance is all things, inasmuch as by its intellect it is able to comprehend all things; whereas every other substance has only a particular participation of being. Consequently God cares for other things for the sake of intellectual substances.

Besides. Whatever happens to a thing in the course of nature happens to it naturally. Now, we see that in the course of nature the intellectual substances use all others for its own sake; either for the perfection of the intellect, which sees the truth in them as in a mirror; or again to sustain the body that is united to an intellectual soul, as is the case in man. It is clear, therefore, that God cares for all things for the sake of intellectual substances.

Moreover. If a man seek something for its own sake, he seeks it always, because what is per se, is always; whereas if he seek a thing on account of something else, he does not of necessity seek it always but only in reference to that for the sake of which he seeks it. Now as we proved above, things derive their being from the divine will. Therefore whatever is always is willed by God for its own sake; and what is not always is willed by God, not for its own sake, but for another's. Now, intellectual substances approach nearest to being always, since they are incorruptible. They are, moreover, unchangeable, except in their choice. Therefore intellectual substances are governed for their own sake, as it were; and others for the sake of intellectual substances.

The fact that all parts of the universe are directed to the perfection of the whole is not in contradiction with the foregoing conclusion; since all the parts are directed to the perfection of the whole, in so far as one part serves another. Thus in the human body it is clear that the lungs belong to the body's perfection, in that they serve the heart; wherefore there is no contradiction in the lungs being for the sake of the heart, and for the sake of the whole animal. In like manner that other natures are on account of the intellectual is not contrary to their being for the perfection of the intellectual substance, the universe would not be complete.

Nor again does the fact that individuals are for the sake of the species militate against what has been said. Because through being directed to their species, they are directed also to the intellectual nature. For a corruptible thing is directed to man, not on account of only one individual man, but on account of the whole human species. Yet a corruptible thing could not serve the whole human species, except as regards its own entire species. Hence the order whereby corruptible things are directed to man, requires that individuals be directed to the species.

When we assert that intellectual substances are directed by divine providence for their own sake, we do not mean that they are not also referred by God and for the perfection of the universe. Accordingly they are said to be provided for on their account, and others on account of them, because the goods bestowed on them by divine providence are not given them for another's profit: whereas those bestowed on others are in the divine plan intended for the use of intellectual substances. Hence it is said (Deut 4:19) *Lest thou see the sun and the moon and the other stars, and being deceived by error, thou adore and serve them, which the Lord thy god created for the service of all the nations that are under heaven: and (Ps. 8:8): Thou hast subjected all things under his feet, all sheep and oxen; moreover, the beasts also of the field: and (Wis. 12:18): thou, being master of power, judgest with tranquility, and the great favour disposest of us.*

Hereby is refuted the error of those who said it is sinful for a man to kill dumb animals: for by divine providence they are intended for man's use in the natural order. Hence it is not wrong for man to make use of the animal, either by killing of or in any other way whatever. For this reason the Lord said to Noah (Gen. 9:3) *As the green herbs I have delivered all flesh to you.*

And if any passages of Holy Writ seem to forbid us to be cruel to dumb animals, for instance to kill a bird with its young: this is either to remove man's thoughts from being cruel to other men, and lest through being cruel to animals one becomes cruel to human beings: or because injury to an animal leads to the temporal hurt of man, either of the doer of the deed, or of another: or on account of some signification: thus the Apostle expounds the prohibition against *muzzling the ox that treadeth the corn.*

Animals Are Machines

RENÉ DESCARTES

If one is to have special privileges, one must first establish some special and distinguishing characteristics that make them different from all others. For example, if I believe that I have the right to cheat on my taxes, I either have to agree that everyone else has that right (which would lead to chaos, an undesirable outcome even for me) or I have to demonstrate how I differ from everyone else, and how that difference then allows me to cheat.

This is the problem faced by the philosophers during the Enlightenment with regard to other creatures. They began to realize that other animals are put together pretty much the same way humans are – bones, ligaments, muscles, blood, and so on. But if non-human animals were to be treated differently, such as eating their flesh or abusing and causing them pain, then there had to be some significant difference between humans and other animals that made such actions morally acceptable. René Descartes (1596–1650) (famous to engineers mostly for discovering what we now call the Cartesian coordinates) addressed this problem in a letter to a friend, part of which is reproduced here.

I had explained all these matters in some detail in the Treatise which I formerly intended to publish. And afterwards I had shown there, what must be the fabric of the nerves and muscles of the human body in order that the animal spirits whether contained should have the power to move the members, just as the heads of animals, a little while after decapitation, are still observed

R. Descartes, "Discourse on the Method," in *The Philosophical Works of Descartes* 1:115–118, translated by Elizabeth S. Haldane and G. R. T. Ross (Cambridge University Press, New York, 1931). Copyright © 1931 by Cambridge University Press. Reprinted with the permission of Cambridge University Press.

to move and bite the earth, notwithstanding that they are no longer animate; what changes are necessary in the brain to cause wakefulness, sleep and dreams; how light, sounds, smells, tastes, heat and other qualities pertaining to external object are able to imprint on it various ideas by the intervention of the senses; how hunger, thirst and other internal affections can also convey their impressions upon it; what should be regarded as the "common sense" by which these ideas are received, and what is meant by the memory which retains them, by the fancy which can change them in diverse ways and out of them constitute new ideas, and which, by the same means, distribute the animal spirits through the muscles, can cause the members of such a body to move in as many diverse ways, and in a manner as suitable to the objects which present themselves to its senses and to its internal passions, as can happen in our own case apart from the direction of our free will. And this will not seem strange to those, who, knowing how many different *automata* or moving machines can be made by the industry of man, without employing in so doing more than a very few parts in comparison with the great multitude of bones, muscles, nerves, arteries, veins, or other parts that are found in the body of each animal. From this aspect the body is regarded as a machine which, having been made by the hands of God, is incomparably better arranged, and possesses in itself movements which are much more admirable, than any of those which can be invented by man. Here I specially stopped to show that if there had been such machines, possessing the organs and outward form of a monkey or some other animal without reason, we should not have had any means of ascertaining that they were not of the same nature as those animals. On the other hand, if there were machines which bore a resemblance to our body and imitated our actions as far as it was morally possible to do so, we should always have two very certain tests by which to recognize that, for all that, they were not real men. The first is, that they could never use speech or other signs as we do when placing our thoughts on record for the benefit of others. For we can easily understand a machine's being constituted so that it can utter works, and even emit some responses to action on it of a corporeal kind, which brings about a change in its organs; for instance, if it is touched in a particular part it may ask what we wish to say to it; if in another it may exclaim that it is being hurt, and so on. But it never happens that it arranges its speech in various ways, in order to reply appropriately to varying that may be said in its presence, as even the lowest type of man can do. And the second deference is, that although machines can perform certain things as well or perhaps better than any of us can, they infallibly fall short in others, by which means we may deduce that they did not act from knowledge, but only from the disposition of their organs. For while reason is a universal instrument which can serve for all contingencies, these organs have need of some special adaptation for every particular action. From thus it follows that it is morally impossible that there should be sufficient diversity in any machine to allow it to act in all the events of life in the same way as our reason causes us to act.

By these two methods we may also recognize the difference that exists between men and brutes. First it is a very remarkable fact that there are none so deprived and stupid, without even excepting idiots, that they cannot arrange different words together, forming of them a statement by which they make known their thought; while, on the other hand, there is no other animal, however perfect and fortunately circumstanced it may be, which can do the same. It is not the want of organs that brings this to pass, for it is evident that magpies and parrots are able to utter words just like ourselves, and yet they cannot speak as we do. That is, so as to give evidence that they think of what they say. On the other hand, men who, being born deaf and dumb, are in the same degree, or even more than the brutes, destitute of the organs which serve the others for talking, are in the habit of themselves inventing certain signs by which they make themselves understood by those who, being usually in their company have leisure to learn their language. And this does not merely show that the brutes have less reason than men, that they have none at all, since it is clear that very little is required in order to be able to talk. And when we notice the inequality that exists between animals of the same species, as well as between men, and observe that some are more capable of receiving instruction than others, it is not credible that a monkey or a parrot, selected as the most perfect of its species, should not in these matters equal the stupidest child to be found, or at least a child whose mind is clouded, unless in the case of the brute the soul were of an entirely different nature from ours. And we ought not to confound speech with natural movements which betray passions and may be imitated by machines as well as be manifested by animals; nor must we think, as did some of the ancients, that brutes talk, although we do not understand their language. For if this were true, since they have many organs which are allied to our own, they could communicate their thoughts to us just as easily as those of their own race. It is also a very remarkable fact that although there are many animals which exhibit more dexterity than we do in some of their actions, we at the same time observe that they do not manifest any dexterity at all in many others. Hence the fact that they do better than we do, does not prove that they are endowed with mind, for in this case they would have more reason than any of us, and should surpass us in all other things. It rather shows that they have no reason at all, and that it is nature which acts in them according to the disposition of their organs, just as a clock, which is only composed of wheels and weights is able to tell the hours and measure the time more correctly than we can do with all our wisdom.

I had described after this the rational soul and shown that it could not in any way derive from the power of matter, like the other things of which I had spoken, but that it must be expressly created. I showed too, that it is not sufficient that it should be lodged in the human body like a pilot in his ship, unless perhaps for the moving of its members, but that it is necessary that it should also be joined and united more closely to the body in order to have sensations and appetites similar to our own, and thus to form a true man. In conclusion, I have here enlarged a little on the subject of the soul, because it is one of the greatest importance. For next to the error of those who deny God, which I think I have already sufficiently refuted, there is none which is more effectual in leading feeble spirits from the straight path of virtue, than to imagine that the soul of the brute is of the same nature as our own, and in consequence, after this life we have nothing to fear or to hope for, any more than the flies and ants. As a matter fact, when one comes to know how greatly they differ, we understand much better the reasons which go to prove that our soul is in its nature entirely independent of body, and in consequence that it is not liable to die with it. And then, inasmuch as we observe no other causes capable of destroying it, we are naturally included to judge that it is immortal.

Why We Have No Obligations to Animals

IMMANUEL KANT

Perhaps no other philosopher has had as much influence on the field of ethics as has Immanuel Kant (1724–1804). He was born in and lived all his life in Königsberg, Prussia, now Kaliningrad (part of Russia). His ethical reasoning and arguments are the most pure form of reasoning and have inspired untold students and writers. His basic premise was that correct ethics demands that every person be treated as an end in his- or herself, and not as a means to an end, and that each of us has a duty to all others to behave ethically. This concept requires reciprocity among humans, and hence there is no room in the argument for animals or inanimate things. In this excerpt Kant discusses the place of animals in his philosophy.

Baumgarten speaks of duties towards beings which are beneath us and beings which are above us. But so far as animals are concerned, we have no direct duties. Animals are not self-conscious and are there merely as a means to an end. That end is man. We can ask, "Why do animals exist?" But to ask, "Why does man exist?" is a meaningless question. Our duties towards animals are merely indirect duties towards humanity. Animal nature has analogies to human nature, and by doing our duties to animals in respect of manifestations which correspond to manifestations of human nature, we indirectly do our duty towards humanity. Thus, if a dog has served his master long and faithfully, his service, on the analogy of human service, deserves reward, and when the dog has grown too old to serve, his master ought to keep him until he dies. Such action helps to support us in our duties towards human beings, where they are bounden duties. If then any acts of animals

Excerpted from *Lectures on Ethics* by Immanuel Kant, translated by Louis Infield (Methuen & Co., London, 1963). Used with permission.

are analogous to human acts and spring from the same principles, we have duties towards the animals because thus we cultivate the corresponding duties towards human beings. If a man shoots his dog because the animal is no longer capable of service, he does not fail in his duty to the dog, for the dog cannot judge, but his act is inhuman and damages in himself that humanity which it is his duty to show towards mankind. If he is not to stifle his human feelings, he must practice kindness towards animals, for he who is cruel to animals becomes hard also in his dealings with men. We can judge the heart of a man by his treatment of animals. Hogarth depicts this in his engravings. He shows how cruelty grows and develops. He shows the child's cruelty to animals, pinching the tail of a dog or a cat; he then depicts the grown man in his cart running over a child; and lastly, the culmination of cruelty in murder. He thus brings home to us in a terrible fashion the rewards of cruelty, and this should be an impressive lesson to children. The more we come in contact with animals and observe their behavior, the more we love them, for we see how great is their care for their young. It is then difficult for us to be cruel in thought even to a wolf. Leibnitz used a tiny worm for purposes of observation, and then carefully replaced it with its leaf on the tree so that it should not come to harm through any act of his. He would have been sorry – a natural feeling for a humane man – to destroy such a creature for no reason. Tender feelings towards dumb animals develop humane feelings toward mankind. In England butchers and doctors do not sit on a jury because they are accustomed to the sight of death and hardened. Vivisectionists, who use living animals for their experiments, certainly act cruelly, although their aim is praiseworthy, and they can justify their cruelty, since animals must be regarded as man's instruments; but any such cruelty for sport cannot be justified. A master who turns out his ass or his dog because the animal can no longer earn its keep manifests a small mind. The Greeks' ideas in this respect were high-minded, as can be seen from the fable of the ass and the bell of ingratitude. Our duties towards animals, then, are indirect duties towards mankind.

The Ethical Relationship Between Humans and Other Organisms

R. D. GUTHRIE

Guthrie, a zoologist living in Alaska, defines in this article the anthropocentric view of classical ethics in the modern context. He reiterates the classical position that ethics is strictly intrahuman, and that other creatures simply cannot be included in ethical thinking. He notes that any species, including humans, cannot incorporate another species in its code of moral conduct. While Descartes and Kant did not appreciate the principles of ecology, Guthrie believes that the well-being of other species is important since this could affect our own well-being, and it therefore would be shortsighted of us to destroy other species or to wreck the environment.

An analysis of one aspect of our ethical system necessarily involves some dealing with the system as a whole. Unfortunately, the area of ethical theory can sometimes be an ideological quagmire from which few return enlightened. Much of the difficulty arises from our being drawn, by tradition, into thinking that our moral scaffolding is suspended from some outside agency. Rather, I would subscribe to the concept that moral principles, and the standards by which they are judged, are human constructs and thus can be evaluated on an empirical basis, even though the criteria are complex and the judgements sometimes difficult. Inherent in this position is the idea that our judgements are dependent upon generalizations from past experiences and may have to be altered as new situations are encountered. The only aprioristic element is the underlying assumption that man's rules of conduct are

R. D. Guthrie, "The Ethical Relationship Betwwen Humans and Other Organisms," *Perspectives in Biology and Medicine*, vol. 11 (1967) pp. 52–62. Used with permission.

to be to his benefit. For the limited purposes of this essay, I will thus assume that the most desirable rules governing human behavior are those which, now and in the future, promote the welfare of human population as an aggregate of individuals and contribute to the smooth functioning of its social machinery, while at the same time allowing for the greatest freedom of individual expression and fulfillment. Such a conceptual distillation necessarily includes academic deficiencies with which philosophers of ethical theory will quibble; but, by and large, this has become the gauge by which we evaluate political system, economic policies, codes of sexual behavior, technological innovations, planned parenthood, and so forth. This assumption of the most desirable code of conduct forms the basis of the idea that I wish to present.

My thesis is that the inclusion of other organisms as primary participants in our ethical system is both logically unsound and operationally unfeasible. It is illogical because we cannot consider other organisms as moral bodies and amoral bodies simultaneously. By *moral bodies* I mean those entities ultimately to be considered in evaluating the action. As an example of this categorization, let us say that, as part of an experiment, a mineralogist wishes to dissolve a unique crystal. On what basis does he decide that the destruction will be worthwhile? The judgement has to be made on the effect of his action on living and future humans – the immediate benefits derived from the crystal's destruction weighed against the assets of its continued existence. The rare crystal is an amoral body, since our concern is with the ultimate effect on humans (the moral bodies which made this a moral question) and not with the welfare of the rock per se.

The relationship among non–human organisms are also not generally defined as moral or immoral. Most would agree that a wild wolf killing a wild deer is, in and of itself, not subject to moral analysis. We as humans may wish to keep the wolf from killing a game species, or the weeds from stunting the turnips, but we do not contend that the wolves and weeds are immoral for so doing. We wish to curtail the wolf and weed population, not ultimately for the sake of the deer and turnips, but for our own ends. In my categorization, then, rare crystals, wolves, deer, weeds, and turnips are all amoral bodies. Likewise, an act of another organism towards humans, say a mosquito bite, could also be classified as amoral. The mosquito is an amoral body, and we do not hold it morally responsible for having bitten us. Thus, at two bounds of triangular relationship, we recognize that amoral nature of the non–human's act toward another non–human organism and, second, a non–human organism's act toward a human as being amoral. It is difficult not to conclude that the final bond – a human's act toward other organisms – is, in and of itself, an amoral one. It becomes a moral act only when humans are affected, because our moral codes are rules of human behavior, as I assumed in the beginning, and as such, exclude other organisms as primary participants.

Supplemental Readings for Chapter 5

The Case for Animal Rights

TOM REGAN

One of the most staunch and most vocal animal rights support-
ers is Tom Regan, professor of philosophy at North Carolina
State University. His book *The Case for Animal Rights* is a classic
in the field. The following excerpt is from a shorter paper in a
well-read anthology edited by Peter Singer.

I regard myself as an advocate of animal rights – as a part of the animal rights
movement. That movement, as I conceive it, is committed to a number of
goals, including:

The total abolition of the use of animals in science;
The total dissolution of commercial animal agriculture;
The total elimination of commercial and sport hunting and trapping.

There are, I know, people who profess to believe in animal rights but do not
avow these goals. Factory farming, they say, is wrong – it violates animals'
rights – but traditional animal agriculture is all right. Toxicity tests of cos-
metics on animals violates their rights, but important medical research –
cancer research, for example – does not. The clubbing of baby seals is abhor-
rent, but not the harvesting of adult seals. I used to think I understood this
reasoning. Not any more. You don't change unjust institutions by tidying
them up.

What's wrong – fundamentally wrong – with the way animals are treated
isn't the details that vary from case to case. It's the whole system. The for-
lornness of the veal calf is pathetic, heart wrenching; the pulsing pain of the
chimp with electrodes planted deep in her brain is repulsive; the slow, tor-
turous death of the raccoon caught in the leg-hold trap is agonizing. But

T. Regan, "The Case for Animal Rights," in Peter Singer, ed., *In Defense of Animals* (Blackwell Publishers, Oxford, 1985). Used with permission.

what is wrong isn't the pain, isn't the suffering, isn't the deprivation. These compound what's wrong. Sometimes – often – they make it much, much worse. But they are not the fundamental wrong.

The fundamental wrong is the system that allows us to view animals as *our resources,* here for *us* – to be eaten, or surgically manipulated, or exploited for sport or money. Once we accept this view of animals – as our resources – the rest is as predictable as it is regrettable. Why worry about their loneliness, their pain, their death? Since animals exist for us, to benefit us in one way or another, what harms them really doesn't matter – or matters only if it starts to bother us, makes us feel a trifle uneasy when we eat our veal escallop, for example. So, yes, let us get veal calves out of solitary confinement, give them more space, a little straw, a few companions. But let us keep our veal escallop.

But a little straw, more space and a few companions won't eliminate – won't even touch – the basic wrong that attaches to our viewing and treating these animals as our resources. A veal calf killed to be eaten after living in close confinement is viewed and treated in this way: but so, too, is another who is raised (as they say) 'more humanely.' To right the wrong of our treatment of farm animals requires more than making rearing methods 'more humane'; it requires the total dissolution of commercial animal agriculture.

Animals, it is true, lack many of the abilities humans possess. They can't read, do higher mathematics, build a bookcase or make baba ghanoush. Neither can many human beings, however, and yet we don't (and shouldn't) say that they (these humans) therefore have less inherent value, less of a right to be treated with respect, than do others. It is the *similarities* between those human beings who most clearly, most non-controversially have such value (the people reading this, for example), not our differences, that matter most. And the really crucial, the basic similarity is simply this: we are each of us the experiencing subject of a life, a conscious creature having an individual welfare that has importance to us whatever our usefulness to others. We want and prefer things, believe and feel things, recall and expect things. And all these dimensions of our life, including our pleasure and pain, our enjoyment and suffering, our satisfaction and frustration, our continued existence or our untimely death – all make a difference to the quality of our life as lived, as experienced, by us as individuals. As the same is true of those animals that concern us (the ones what are eaten and trapped, for example), they too must be viewed as the experiencing subjects of a life, with inherent values of their own.

Some may resist the idea that animals have inherent value. "Only humans have such value," they profess. How might this narrow view be defended? Shall we say that only humans have the requisite intelligence, or autonomy, or reason? But there are many, many humans who fail to meet these standards and yet are reasonably viewed as having value above and beyond their usefulness to others. Shall we claim that only humans belong to the right species, the species *Homo sapiens?* But this is blatant speciesism. Will it be said, then, that all – and only – humans have immortal souls? Then our opponents have their work cut out for them. I am myself not ill-disposed to the proposition that there are immortal souls. Personally, I profoundly hope I have one. But I would not want to rest my position on a controversial ethical issue on the even more controversial question about who or what has an immortal soul. That is to dig one's hole deeper, not to climb out. Rationally, it is better to resolve moral issues without making more controversial assumptions than are needed. The question of who has inherent value is such a question, one that is resolved more rationally without the introduction of the idea of immortal souls than by its use.

Well, perhaps some will say that animals have some inherent value, only less than we have. Once again, however, attempts to defend this view can be shown to lack rational justification. What could be the basis of our having more inherent value than animals? Their lack of reason, or autonomy, or intellect? Only if we are willing to make the same judgement in the case of humans who are similarly deficient. But it is not true that such humans – the retarded child, for example, or the mentally deranged – have less inherent value than you or I. Neither, then, can we rationally sustain the view that animals like them in being the experiencing subjects of a life have less inherent value. *All* who have inherent value have it *equally*, whether they be human animals or not.

Inherent value, then, belongs equally to those who are the experiencing subjects of a life. Whether it belongs to others – to rocks and rivers, trees and glaciers, for example – we do not know and may never know. But neither do we need to know, if we are to make the case for animal rights. We do not need to know, for example, how many people are eligible to vote in the next presidential election before we can know whether I am. Similarly, we do not need to know how many individuals have inherent value before we can know that some do. When it comes to the case for animal rights, then, what we need to know is whether the animals that, in our culture, are routinely eaten, hunted and used in our laboratories, for example, are like us in being subjects of a life. And we do know this. We do know that many – literally, billions and billions – of these animals are the subjects of a life in the sense explained and so have inherent value if we do. And since, in order to arrive at the best theory of our duties to one another, we must recognize our equal inherent value as individuals, reason – not sentiment, not emotion – reason compels us to recognize the equal inherent value of these animals, and, with this, their equal right to be treated with respect.

The Land Ethic

ALDO LEOPOLD

Aldo Leopold (1887–1948), a naturalist and writer, produced perhaps the first formal statement on an environmental ethic. His *A Sand County Almanac* is full of touching and vivid observations of the natural environment. He was a realist who recognized that the preservationists' ideas could never be implemented in their pure form; yet he also rejected the ideas of the early conservationists, who valued nature only for what it could provide in the form of material wealth. He insisted that nature has an intrinsic value beyond that of dollars and cents, and that this value could not necessarily be expressed in terms of a benefit–cost ratio.

In this short excerpt from his book, the idea of the land ethic is first introduced. This concept was eventually developed into what has become the "ecocentric" approach to environmental ethics. Bear in mind in reading this article that it was written in the 1930s, and that the idea of ethics applied to land, thus widening the moral community, was a revolutionary concept.

When god-like Odysseus returned from the wars in Troy, he hanged all on one rope a dozen slave-girls of his household whom he suspected of misbehavior during his absence.

This hanging involved no question of propriety. The girls were property. The disposal of property was then, as now, a matter of expediency, not of right and wrong.

Excerpted from A. Leopold, *A Sand County Almanac* (Oxford University Press, New York, 1966; originally published in 1949). Copyright 1966 by Aldo Leopold. Used by permission of Oxford University Press, Inc.

Concepts of right and wrong were not lacking from Odysseus' Greece: witness the fidelity of his wife through the long years before at last his black-prowed galleys clove the wine-dark seas for home. The ethical structure of that day covered wives, but had not yet been extended to human chattels. During the three thousand years which have since elapsed, ethical criteria have been extended to many fields of conduct, with corresponding shrinkages in those judged by expediency only.

The Ethical Sequence

This extension of ethics, so far studied only by philosophers, is actually a process in ecological evolution. Its sequences may be described in ecological as well as in philosophical terms. An ethic, ecologically, is a limitation on freedom of action in the struggle for existence. An ethic, philosophically, is a differentiation of social from anti-social conduct. These are two definitions of one thing. The thing has its origin in the tendency of interdependent individuals or groups to evolve modes of co-operation. The ecologist calls these symbioses. Politics and economics are advanced symbioses in which the original free-for-all competition has been replaced, in part, by co-operative mechanisms with an ethical content.

The complexity of co-operative mechanisms has increased with population density, and with the efficiency of tools. It was simpler, for example, to define the anti-social uses of sticks and stones in the days of the mastodons than of bullets and billboards in the age of motors.

The first ethics dealt with the relation between individuals; the Mosaic Decalogue is an example. Later accretions dealt with the relation between the individual and society. The Golden Rule tries to integrate the individual to society; democracy to integrate social organization to the individual.

There is as yet no ethic dealing with man's relation to land and to the animals and plants which grow upon it. Land, like Odysseus' slave girls, is still property. The land-relation is still strictly economic, entailing privileges but not obligations.

The extension of ethics to this third element in human environment is, if I read the evidence correctly, an evolutionary possibility and an ecological necessity. It is the third step in a sequence. The first two have already been taken. Individual thinkers since the days of Ezekiel and Isaiah have asserted that the despoliation of land is not only inexpedient but wrong. Society, however, has not yet affirmed their belief. I regard the present conservation movement as the embryo of such an affirmation.

An ethic may be regarded as a mode of guidance for meeting ecological situations so new or intricate or involving such deferred reactions, that the path of social expediency is not discernible to the average individual. Animal instincts are modes of guidance for the individual in meeting such situations. Ethics are possibly a kind of community instinct in-the-making.

The Community Concept

All ethics so far evolved rest upon a single premise: that the individual is a member of a community of interdependent parts. His instincts prompt him to compete for his place in the community, but his ethics prompt him also to co-operate (perhaps in order that there may be a place to compete for).

The land ethic simply enlarges the boundaries of the community to include soils, waters, plants, and animals, or collectively: the land.

This sounds simple: do we not already sing our love for and obligation to the land of the free and the home of the brave? Yes, but just what and whom do we love? Certainly not the soil, which we are sending helter-skelter downriver. Certainly not the waters, which we assume have no function except to turn turbines, float barges, and carry off sewage. Certainly not the plants, of which we exterminate whole communities without batting an eye. Certainly not the animals, of which we have already extirpated many of the largest and most beautiful species. A land ethic of course cannot prevent the alteration, management, and use of these 'resources,' but it does affirm their right to continued existence, and at least in spots, their continued existence in a natural state.

In short, a land ethic changes the role of *Homo sapiens* from conqueror of the land–community to plain member and citizen of it. It implies respect for his fellow-members, and also respect for the community as such.

In human history, we have learned (I hope) that the conqueror role is eventually self-defeating. Why? Because it is implicit in such a role that the conqueror knows, *ex cathedra,* just what makes the community clock tick, and just what and who is valuable, and what and who is worthless, in community life. It always turns out that he knows neither, and this is why his conquests eventually defeat themselves.

In the biotic community, a parallel situation exists. Abraham knew exactly what the land was for: it was to drip milk and honey into Abraham's mouth. At the present moment, the assurance with which we regard this assumption is inverse to the degree of our education.

The ordinary citizen today assumes that science knows what makes the community clock tick; the scientist is equally sure that he does not. He knows that the biotic mechanism is so complex that its workings may never be fully understood.

That man is, in fact, only a member of a biotic team is shown by an ecological interpretation of history. Many historical events, hitherto explained solely in terms of human enterprise, were actually biotic interactions between people and land. The characteristics of the land determined the facts quite as potently as the characteristics of the men who lived on it.

Consider, for example, the settlement of the Mississippi valley. In the years following the Revolution, three groups were contending for its control: the native Indian, the French and English traders, and the American settlers. Historians wonder what would have happened if the English at Detroit had thrown a little more weight into the Indian side of those tipsy scales which decided the outcome of the colonial migration into the cane-lands of Kentucky. It is time now to ponder the fact that the cane-lands, when subjected to the particular mixture of forces represented by the cow, plow, fire, and axe of the pioneer, became bluegrass. What if the plant succession inherent in this dark and bloody ground had, under the impact of these forces, given us some worthless sedge, shrub, or weed? Would Boone and Kenton have held out? Would there have been any overflow into Ohio, Indiana, Illinois, and Missouri? Any Louisiana Purchase? Any transactional union of new states? Any Civil War?

Kentucky was one sentence in the drama of history. We are commonly told what the human actors in this drama tried to do, but we are seldom told that their success, or the lack of it, hung in large degree on the reaction of particular soils to the impact of the particular forces exerted by their occupancy. In the case of Kentucky, we do not even know where the bluegrass came from – whether it is a native species, or a stowaway from Europe.

Contrast the cane-lands with what hindsight tells us about the Southwest, where pioneers were equally brave, resourceful, and persevering. The impact of occupancy here brought no bluegrass, or other plant fitted to withstand the bumps and buffetings of hard use. This region, when grazed by livestock, reverted through a series of more and more worthless grasses, shrubs, and weeds to a condition of unstable equilibrium. Each recession of plant types bred erosion; each increment to erosion bred a further recession of plants. The result today is a progressive and mutual deterioration, not only of plants and soils, but of the animal community subsisting thereon. The early settlers did not expect this: on the ciénegas of New Mexico some even cut ditches to hasten it. So subtle has been its progress that few residents of the region are aware of it. It is quite invisible to the tourist who finds this wrecked landscape colorful and charming (as indeed it is, but it bears scant resemblance to what it was in 1848).

This same landscape was 'developed' once before, but with quite different results. The Pueblo Indians settled the Southwest in pre-Columbian times, but they happened *not* to be equipped with range livestock. Their civilization expired, but not because their land expired.

In India, regions devoid of any sod-farming grass have been settled, apparently without wrecking the land, by the simple expedient of carrying the grass to the cow, rather than vice versa. (Was this the result of some deep wisdom, or was it just good luck? I do not know.)

In short, the plant succession steered the course of history; the pioneer simply demonstrated, for good or ill, what successions inhered in the land. Is history taught in this spirit? It will be, once the concept of land as a community really penetrates our intellectual life.

The Ecological Conscience

Conservation is a state of harmony between men and land. Despite nearly a century of propaganda, conservation still proceeds at a snail's pace; progress still consists largely of letterhead pieties and convention oratory. On the back forty we still slip two steps backward for each forward stride.

The usual answer to this dilemma is 'more conservation education.' No one will debate this, but is it certain that only the *volume* of education needs stepping up? Is something lacking in the *content* as well?

It is difficult to give a fair summary of its content in brief form, but, as I understand it, the content is substantially this: obey the law, vote right, join some organizations, and practice what conservation is profitable on your own land; the government will do the rest.

Is not this formula too easy to accomplish anything worth-while? It defines no right or wrong, assigns no obligation, calls for no sacrifice, implies no change in the current philosophy of values. In respect of land-use, it urges only enlightened self-interest. Just how far will such education take us? An example will perhaps yield a partial answer.

By 1930 it had become clear to all except the ecologically blind that south-western Wisconsin's topsoil was slipping seaward. In 1933 the farmers were told that if they would adopt certain remedial practices for five years, the public would donate CCC labor to install them, plus necessary machinery and materials. The offer was widely accepted, but the practices were widely forgotten when the five-year contract period was up. The farmers continued only those practices that yielded an immediate and visible economic gain for themselves.

This led to the idea that maybe farmers would learn more quickly if they themselves wrote the rules. Accordingly the Wisconsin Legislature in 1937 passed the Soil Conservation District Law. This said to farmers, in effect: *We, the public, will furnish you free technical service and loan you specialized machinery, if you will write your own rules for land-use. Each county may write its own rules, and these will have the force of law.* Nearly all the counties promptly organized to accept the proffered help, but after a decade of oper-ation, *no county has yet written a single rule.* There has been visible progress in such practices as strip–cropping, pasture renovation, and soil liming, but none in fencing woodlots against grazing, and none in excluding plow and cow from steep slopes. The farmers, in short, have selected those remedial practices which were profitable anyhow, and ignored those which were prof-itable to the community, but not clearly profitable to themselves.

When one asks why no rules have been written one is told that the com-munity is not yet ready to support them; education must precede rules. But the education actually in progress makes no mention of obligations to land over and above those dictated by self-interest. The net result is that we have more education but less soil, fewer healthy woods, and as many floods as in 1937.

The puzzling aspect of such situations is that the existence of obligations over and above self-interest is taken for granted in such rural community enterprises as the betterment of roads, schools, churches, and baseball teams. Their existence is not taken for granted, nor as yet seriously dis-cussed, in bettering the behavior of the water that falls on the land, or in the preserving of the beauty or diversity of the farm landscape. Land-use ethics are still governed wholly by economic self-interest, just as social ethics were a century ago.

To sum up: we asked the farmer to do what he conveniently could to save his soil, and he has done just that, and only that. The farmer who clears the woods off a 75 per cent slope, turns his cows into the clearing, and dumps its rainfall, rocks, and soil into the community creek, is still (if otherwise decent) a respected member of society. If he puts lime on his fields and plants his crops on contour, he is still entitled to all the privileges and emoluments of his Soil Conservation District. The District is a beautiful piece of social machinery, but it is coughing along on two cylinders because we have been too timid, and too anxious for quick success, to tell the farmer the true magnitude of his obligations. Obligations have no meaning without conscience, and the problem we face is the extension of the social conscience from people to land.

No important change in ethics was ever accomplished without an internal change in our intellectual emphasis, loyalties, affections, and convictions. The proof that conservation has not yet touched these foundations of conduct lies in the fact that philosophy and religion have not yet heard of it. In our attempt to make conservation easy, we have made it trivial.

Substitutes for a Land Ethic

When the logic of history hungers for bread and we hand out a stone, we are at pains to explain how much the stone resembles bread. I now describe some of the stones which serve in lieu of a land ethic.

One basic weakness is a conservation system based wholly on economic value. Wildflowers and songbirds are examples. Of the 22,000 higher plants and animals native to Wisconsin, it is doubtful whether more than 5 per cent can be sold, fed, eaten, or otherwise put to economic use. Yet these creatures are members of the biotic community, and if (as I believe) its stability depends on its integrity, they are entitled to continuance.

When one of these non-economic categories is threatened, and if we happen to love it, we invent subterfuges to give it economic importance. At the beginning of the century songbirds were supposed to be disappearing. Ornithologists jumped to the rescue with some distinctly shaky evidence to the effect that insects would eat us up if birds failed to control them. The evidence had to be economic in order to be valid.

It is painful to read these circumlocutions today. We have no land ethic yet, but we have at least drawn nearer the point of admitting that birds should continue as a matter of biotic right, regardless of the presence or absence of economic advantage to us.

A parallel situation exists in respect of predatory mammals, raptorial birds, and fish-eating birds. Time was when biologists somewhat overworked the evidence that these creatures preserve the health of game by killing weaklings, or that they control rodents for the farmer, or that they prey only on "worthless" species. Here again, the evidence had to be economic in order to be valid. It is only in recent years that we hear the more honest argument that predators are members of the community, and that no special interest has the right to exterminate them for the sake of a benefit, real or fancied, to itself. Unfortunately this enlightened view is still in the talk stage. In the field the extermination of predators goes merrily on: witness the impending erasure of the timber wolf by fiat of Congress, the Conservation Bureaus, and many state legislatures.

Some species of trees have been 'read out of the party' by economics-minded foresters because they grow too slowly, or have too low a sale value to pay as timber crops: white cedar, tamarack, cypress, beech, and hemlock are examples. In Europe, where forestry is ecologically more advanced, the non-commercial tree species are recognized as members of the native forest community, to be preserved as such, within reason. Moreover some (like beech) have been found to have a valuable function in building up soil fertility. The interdependence of the forest and its constituent tree species, ground flora, and fauna is taken for granted.

Lack of economic value is sometimes a character not only of species or groups, but of entire biotic communities: marshes, bogs, dunes and 'deserts' are examples. Our formula in such cases is to relegate their conservation to government as refuges, monuments, or parks. The difficulty is that these communities are usually interspersed with more valuable private lands; the government cannot possibly own or control such scattered parcels. The net effect is that we have relegated some of them to ultimate extinction over large areas. If the private owner were ecologically minded, he would be proud to be the custodian of a reasonable proportion of such areas, which add diversity and beauty to his farm and to his community.

In some instances, the assumed lack of profit in these 'waste' areas has proved to be wrong, but only after most of them had been done away with. The present scramble to reflood muskrat marshes is a case in point.

There is a clear tendency in American conservation to relegate to government all necessary jobs that private landowners fail to perform. Government ownership, operation, subsidy, or regulation is now widely prevalent in forestry, range management, soil and watershed management, park and wilderness conservation, fisheries management, and migratory bird management, with more to come. Most of this growth in governmental conservation is proper and logical, some of it is inevitable. That I imply no disapproval of it is implicit in the fact that I have spent most of my life working for it. Nevertheless the question arises: What is the ultimate magnitude of the enterprise? Will the tax base carry its eventual ramifications? At what point will governmental conservation, like the mastodon, become handicapped by its own dimensions? The answer, if there is any, seems to be in a land ethic, or some other force which assigns more obligation to the private landowner.

Industrial landowners and users, especially lumbermen and stockmen, are inclined to wail long and loudly about the extension of government ownership and regulation to land, but (with notable exceptions) they show little disposition to develop the only visible alternative: the voluntary practice of conservation on their own lands.

When the private landowner is asked to perform some unprofitable act for the good of the community, he today assents only with outstretched palm. If the act costs him cash this is fair and proper, but when it costs only forethought, open-mindedness, or time, the issue is at least debatable. The overwhelming growth of land-use subsidies in recent years must be ascribed, in large part, to the government's own agencies for conservation education: the land bureaus, the agricultural colleges, and the extension services. As far as I can detect, no ethical obligation toward land is taught in these institutions.

To sum up: a system of conservation based solely on economic self-interest is hopelessly lopsided. It tends to ignore, and thus eventually to eliminate, many elements in the land community that lack commercial value, but that are (as far as we know) essential to its healthy functioning. It assumes, falsely, I think, that the economic parts of the biotic clock will function without the uneconomic parts. It tends to relegate to government many functions eventually too large, too complex, or too widely dispersed to be performed by government.

An ethical obligation on the part of the private owner is the only visible remedy for these situations.

Reverence for Life

ALBERT SCHWEITZER

Albert Schweitzer (1875–1965) was an amazing person. Born in Germany, he become an organist, physician, scholar, missionary, and philosopher. He spent much of his life in Africa, where he ran a small mission for the benefit of the local people. On one of his trips to Africa, he had a sudden insight, putting into words the philosophy he had been living and espousing. This insight came to him one day while traveling on a boat in Africa. As he describes it:

> At sunset of the third day, near the village of Igendja, we moved along an island in the middle of the wide river. On a sandbank to our left, four hippopotamuses and their young plodded along in our same direction. Just then, in my great tiredness and discouragement, the phrase "Reverence for Life" struck me like a flash. As far as I know, it was a phrase I had never heard nor ever read. I realized at once that it carried within itself the solution to the problem that had been torturing me. Now I know that a system of values which concerns itself only with our relations to other people is incomplete and therefor lacking power for good. Only by means of reverence for life can we establish a spiritual and humane relationship with both people and all living creatures with our reach. Only in this fashion can we avoid harming others, and, within the limits of our capacity, go to their aid whenever they need us.

What follows is an amplification of his "reverence for life" philosophy.

Excerpted from Albert Schweitzer, *Civilization and Ethics*, translated by A. Nash (Black, London, 1923). Used with permission.

True philosophy must commence with the most immediate and comprehensive fact of consciousness. And this may be formulated as follows: "I am life which wills to live, and I exist in the midst of life which wills to live." This is no more excogitated subtlety. Day after day and hour after hour I proceed on my way invested in it. In every moment of reflection it forces itself on me anew. A living world- and life-view, informing all the facts of life, gushes forth from it continually, as from an eternal spring. A mystical ethical oneness with existence grows forth from it unceasingly. . . .

Ethics thus consists of this, that I experience the necessity of practicing the same reverence for life toward all will-to-live, as toward my own. Therein I have already the needed fundamental principle of morality. It is *good* to maintain and cherish life; it is *evil* to destroy and to check life.

As a matter of fact, everything which is the usual ethical valuation of inter-human relations is looked upon as good and can be traced back to the material and spiritual maintenance or enhancement of human life and to the effort to raise it to its highest level of value. And contrariwise everything in human relations which is considered as evil, is in the final analysis found to be material or spiritual destruction or checking of human life and slackening of the effort to raise it to its highest value. Individual concepts of good and evil which are widely divergent and apparently unconnected fit into one another like pieces which belong together, the moment they are comprehended and their essential nature is grasped in this general nation.

The fundamental principle of morality which we seek as a necessity for thought is not, however, a matter only of arranging and deepening correct views of good and evil, but also of expanding and extending these. A man is really ethical only when he obeys the constraint laid on him to help all life which he is able to succor, and when he goes out of his way to avoid injuring anything living. He does not ask how far this or that life deserves sympathy as valuable in itself, nor how far it is capable of feeling. To him life as such is sacred. He shatters no ice crystal that sparkles in the sun, tears no leaf from its tree, breaks off no flower, and is careful not to crush any insect as he walks. If he works by lamplight on a summer evening, he prefers to keep the window shut and to breathe stifling air, rather than to see insect after insect fall on his table with singed and sinking wings.

If he goes out into the street after a rainstorm and sees a worm which has strayed there, he reflects that it will certainly dry up in the sunshine, if it does not quickly regain the damp soil into which it can creep and so he helps it back from the deadly paving stones into the lush grass. Should he pass by an insect which has fallen into a pool, he spares the time to reach it a leaf or stalk on which it may clamber and save itself.

He is not afraid of being laughed at as sentimental. It is indeed the fate of every truth to be an object of ridicule when it is first acclaimed. It was once considered foolish to suppose that coloured men were really human beings and ought to be treated as such. What was once foolishness has now become a recognized truth. Today it is considered as exaggeration to proclaim constant respect for every form of life as being the serious demand of a rational ethic. But the time is coming when people will be amazed that the human race was so long before it recognized that thoughtless injury to life is incompatible with real ethics. Ethics is in its unqualified form extended responsibility with regard to everything that has life.

The Ethics of Respect for Nature

PAUL TAYLOR

Paul Taylor is the most effective writer for what he calls the "bio-centric" view of environmental ethics. In this excerpt from his paper "The Ethics of Respect for Nature," he states his main theses for his arguments for extending the moral community to include all living things, which eventually constituted his major book, *Respect for Nature.*

Taylor is professor of philosophy at Brooklyn College in New York.

What would justify acceptance of a life-centered system of ethical principles? In order to answer this it is first necessary to make clear the fundamental moral attitude that underlies and makes intelligible the commitment to live by such a system. It is then necessary to examine the consideration that would justify any rational agent's adopting that moral attitude.

Two concepts are essential to the taking of a moral attitude of the sort in question. A being which does not "have" these concepts, this is, which is unable to grasp their meaning and conditions of applicability, cannot be said to have the attitude as part of its moral outlook. These concepts are, first, that of the good (well-being, welfare) of a living thing, and second, the idea of an entity possessing inherent worth. I examine each concept in turn.

1. Every organism, species population, and community of life has a good of its own which moral agents can intentionally further or damage by their actions. To say that an entity has a good of its own is simply to say that, without reference to any *other* entity, it can be benefited or harmed. One can act in its overall interest or contrary to its overall interest, and environmental

Paul Taylor, "The Ethics of Respect for Nature," *Environmental Ethics,* vol. 3 (Fall 1981). Used with permission.

conditions can be good for it (advantageous to it) of bad for it (disadvantageous to it). What is bad for an entity is something that is detrimental to its life and well-being.

We can think of the good of an individual nonhuman organism as consisting in the full development of its biological powers. Its good is realized to the extent that it is strong and healthy. It possesses whatever capacities it needs for successfully coping with its environment and so preserving its existence throughout the various stages of the normal life cycle of its species. The good of a population or community of such individuals consists in the population or community maintaining itself from generation to generation as a coherent system of genetically and ecologically related organisms whose average good is at an optimum level for the given environment. (Here *average good* means that the degree of realization of the good of *individual organisms* in the population or community is, on average, greater than would be the case under any other ecologically functioning order of interrelations among those species populations in the given ecosystem.)

The idea of a being having a good of its own, as I understand it, does not entail that the being must have interests or take an interest in what affects its life for better or for worse. We can act in a being's interest or contrary to its interest without its being interested in what we are doing to it in the sense of wanting or not wanting us to do it. It may, indeed, be wholly unaware that favorable and unfavorable events are taking place in its life. I take it that trees, for example, have no knowledge of desires or feelings. Yet it is undoubtedly the case that trees can be harmed or benefited by our actions. We can crush their roots by running a bulldozer too close to them. We can see to it that they get adequate nourishment and moisture by fertilizing and watering the soil around them. Thus we can help or hinder them in the realization of their good. It is the good of trees themselves that is thereby affected. We can similarly act so as to further the good of an entire tree population of a certain species (say, all the redwood trees in a California valley) of the good of a whole community of plant life in a given wilderness area, just as we can do harm to such a population community. . . .

2. The second concept essential to the moral attitude of respect for nature is the idea of inherent worth. We take that attitude toward wild living things (individuals, species populations, or whole biotic communities) when and only when we regard them as entities possessing inherent worth. Indeed, it is only because they are conceived in this way that moral agents can think of themselves as having validly binding duties, obligations, and responsibilities that are *owed* to them as their *due*. I am not at this juncture arguing that they

should be so regarded; I consider it at length below. But so regarding them is a presupposition of our taking the attitude of respect toward them and accordingly understating ourselves as bearing certain moral relations to them. This can be shown as follows.

What does it mean to regard an entity that has a good of its own as possessing inherent worth? Two general principles are involved; the principle of moral consideration and the principle of intrinsic value.

According to the principle of moral consideration, wild living things are deserving of the concern and consideration of all moral agents simply by virtue of their being members of the Earth's community of life. From the moral point of view their good must be taken into account whenever it is affected for better or worse by the conduct of rational agents. This holds no matter what species the creature belongs to. The good of each is to be accorded some value and so acknowledged as having some weight in the deliberations of all rational agents. Of course, it may be necessary for such agents to act in ways contrary to the good of this or that particular organism or group of organisms in order to further the good of others, including the good of humans. But the principle of moral consideration prescribes that, with respect to each being an entity having its own good, every individual is deserving of consideration.

The principle of intrinsic value states that, regardless of what kind of entity it is in other respects, if it is a member of the Earth's community of life, the realization of its good is something *intrinsically* valuable. This means that its good is prima facie worthy of being preserved or promoted as an end in itself and for the sake of the entity whose good it is. Insofar as we regard any organism, species population, or life community as an entity having inherent worth, we believe that it must never be treated as if it were a mere object or thing whose entire value lies in being instrumental to the good of some other entity. The well-being of each is judged to have value in and of itself.

Combining these two principles, we can now define what it means for a living thing or group of living things to possess inherent worth. To say that it possesses inherent worth is to say that its good is deserving of the concern and consideration of all moral agents, and that the realization of its good has intrinsic value, to be pursued as an end in itself and for the sake of the entity whose good it is.

The duties owed to wild organisms, species populations, and communities of life in the Earth's natural ecosystems are grounded on their inherent worth. When rational, autonomous agents regard such entities as possessing inherent worth, they place intrinsic value on the realization of their good and so hold themselves responsible for performing actions that will have this effect and for refraining from actions having the contrary effect.

Why should moral agents regard wild living things in the natural world as possessing inherent worth? To answer this question we must first take into account the fact that, when rational, autonomous agents subscribe to the principles of moral consideration and intrinsic value and so conceive of wild living things as having that kind of worth, such agents are *adopting a certain ultimate moral attitude toward the natural world.* This is the attitude I call "respect for nature." It parallels the attitude of respect for persons in human ethics. When we adopt the attitude of respect for persons as the proper (fitting, appropriate) attitude to take toward all persons as persons, we consider the fulfillment of the basic interests of each individual to have intrinsic value. We thereby make a moral commitment to live a certain kind of life in relation to other persons. We place ourselves under the direction of a system of standards and rules that we consider validly binding on all moral agents as such.

Similarly, when we adopt the attitude of respect for nature as an ultimate moral attitude we make a commitment to live by certain normative principles. These principles constitute the rules of conduct and standards of character that are to govern our treatment of the natural world. This is, first, an *ultimate* commitment because it is not derived from any higher norm. The attitude of respect for nature is not grounded on some other, more general or more fundamental attitude. It sets the total framework for our responsibilities toward the natural world. It can be justified, as I show below, but its justification cannot consist in referring to a more general attitude or more basic normative principle.

Second, the commitment is a *moral* one because it is understood to be a disinterested matter of principle. It is this feature that distinguishes the attitude of respect for nature from the set of feelings and dispositions that comprise the love of nature. The latter stems from one's personal interest in and response to the natural world. Like the affectionate feelings we have toward certain individual human beings, one's love of nature is nothing more than the particular way one feels about the natural environment and its wild inhabitants. And just as our love for an individual person differs from our respect for all persons as such (whether we happen to love them of not), so love of nature differs from respect for nature. Respect for nature is an attitude we believe all moral agents ought to have simply as moral agents, regardless of whether or not they also love nature. Indeed, we have not truly taken the attitude of respect for nature ourselves unless we believe this. To put it in a Kantian way, to adopt the attitude of respect for nature is to take a stance that one wills it to be a universal law for all rational beings. It is to hold that stance categorically, as being validly applicable to every moral agent without exception, irrespective of whatever personal feelings toward nature such an agent might have or might lack.

Although the attitude of respect for nature is in this sense a disinterested and universalizable attitude, anyone who does adopt it has certain steady, more or less permanent depositions. These dispositions, which are themselves to be considered disinterested and universalizable, comprise three interlocking sets; dispositions to seek certain ends, depositions to carry on one's practical reasoning and deliberation in a certain way, and depositions to have certain feelings. We may accordingly analyze the attitude of respect for nature into the following components: (a) The disposition to aim at, and to take steps to bring about, as final and disinterested ends, the promoting and protecting of the good of organisms, species populations, and life communities in natural ecosystems. (These ends are "final" in not being pursued as means to further ends. They are "disinterested" in being independent of the self-interest of the agent.) (b) The disposition to consider actions that tend to realize those ends to be prima facie obligatory *because* they have that tendency. (c) The disposition to experience positive and negative feelings toward states of affairs in the world *because* they are favorable or unfavorable to the good of organisms, species populations, and life communities in natural environments.

The logical connect between the attitude of respect for nature and the duties of a life-centered system of environmental ethics can now be made clear. Insofar as one sincerely takes that attitude and so has the three sets of dispositions, one will at the same time be disposed to comply with certain rules of duty (such as nonmaleficence and noninterference) and with standards of character (such as fairness and benevolence) that determine the obligations one performs and the character traits one develops in fulfilling the moral requirements are the way one *expresses* or *embodies* the attitude in one's conduct and character. . . . I hold that the rules of duty governing our treatment of the natural world and its inhabitants are forms of conduct in which the attitude of respect for nature is manifested.

Should Trees Have Standing?

CHRISTOPHER D. STONE

The Mineral King Valley, in the Sierra Nevada Mountains, is an area of unsurpassed beauty. Adjacent to the Sequoia National Park, the area has been used for limited recreational purposes. The United States Forest Service decided that this area was ripe for more intense recreational use, and in 1965 decided to seek bids for its development. The Walt Disney company was selected, and they proceeded to plan for a large complex of motels, restaurants, and recreational facilities to accommodate 14,000 visitors daily. The Sierra Club reacted with horror to what they considered the despoilment of a pristine wilderness. Unable to persuade the Forest Service to change its mind, the Sierra Club filed a suit in the United States District court to kill the project.

Of greatest issue was whether the Sierra club had "standing to sue," a legal requirement in which the plaintiff has to demonstrate that some wrong has befallen them. In attempt to argue that indeed the Sierra Club has "standing," Christopher Stone, an environmentalist and lawyer, wrote an essay and eventually a book entitled *Should Trees Have Standing?* in order to try to influence the Supreme Court, which would eventually make the final decision.

In the introduction to this essay, Garrett Hardin writes: "Law, to be stable, must be based on ethics. In evoking a new ethic to protect land and other natural amenities, [Aldo] Leopold implicitly called for concomitant changes in the philosophy of the law. Now, less than a generation after the publication of Leopold's

This is the Introduction to Christopher D. Stone, *Should Trees Have Standing?* (William Kaufmann, Los Altos, CA, 1974). Reprinted in *Should Trees Have Standing? And Other Essays on Law, Morals and the Environment* (Oceana Publications, Inc., 1996). Used with permission.

classic essay, Professor Christopher Stone has laid the foundation for just such a philosophy in a graceful essay that itself bids fair to become a classic."

Reproduced is the Introduction to the essay, in which the basic framework for the argument is established. The essay was first written for *Southern California Law Review.*

In *The Descent of Man,* written a full century ago, Charles Darwin observed that the history of man's moral development has been a continual extension in the range of objects receiving his *social instincts and sympathies.* Originally each man had moral concern only for himself and those of a very narrow circle about him; later, he came to regard more and more "not only the welfare, but the happiness of all his fellow men." Then, gradually, "his sympathies became more tender and widely diffused, extending to men of all races, to the imbecile, maimed and other useless members of society, and finally to the lower animals. . . . "

The history of the law suggests a parallel development. The scope of "things" accorded legal protection has been continuously extending. Members of the earliest "families" (including extended kinship groups and clans) treated everyone on the outside as suspect, alien, and rightless, except in the vacant sense of each man's "right to self-defense." "An Indian Thug," it has been written, "conscientiously regretted that he had not robbed and strangled as many travelers as did his father before him. In a rude state of civilization the robbery of strangers is, indeed, generally considered as honorable." And even within a single family, persons we presently regard as the natural holders of at least some legal rights had none. Take, for example, children. We know something of the early rights-status of children from the widespread practice of infanticide – especially of the deformed and female. (Senicide, practiced by the North American Indians, was the corresponding rightlessness of the aged.) Sir Henry Maine tells us that as late as the *Patria Potestas* of the Romans, the father had *jus vitae necisque* – the power of life and death – over his children. It followed legally, Maine writes, that

> he had power of uncontrolled corporal chastisement; he can modify their personal condition at pleasure; he can give a wife to his son; he can give his daughter in marriage; he can divorce his children of either sex; he can transfer them to another family by adoption; and he can sell them.

The child was less than a person: it was, in the eyes of the law, an object, a thing.

The legal rights of children have long since been recognized in principle, and are still expanding in practice. Witness, just within recent time, *In re Gault,* the United States Supreme Court decision guaranteeing basic constitutional protection to juvenile defendants, and the Voting Rights Act of 1970, with its lowering of the voting age to eighteen. We have been making persons of children although they were not, in law, always so. And we have done the same, albeit imperfectly some would say, with prisoners, aliens, women (married women, especially, were nonpersons through most of legal history), the insane, blacks, fetuses, and Indians.

People are apt to suppose that there are natural limits on how far the law can go, that it is only matter in human form that can come to be recognized as the possessor of rights. But it simply is not so. The world of the lawyer is peopled with inanimate right-holders: trusts, corporations, joint ventures, municipalities, Subchapter R partnerships, and nation-states, to mention just a few. Ships, still referred to by courts in the feminine gender, have long had an independent jural life, with striking consequences. In one famous U.S. Supreme Court case a ship had been seized and used by pirates. After the ship's capture, the owners asked for her return; after all, the vessel had been pressed into piracy without their knowledge or consent. But the United States condemned and sold the "offending vessel." In denying release to the owners, Justice Story quoted Chief Justice Marshall from an earlier case:

> This is not a proceeding against the owner; it is a proceeding against the vessel for an offense committed by the vessel; which is not the less an offense . . . because it was committed without the authority and against the will of the owner.

The ship was, in the eyes of the law, the guilty person.

We have become so accustomed to the idea of a corporation having "its" own rights, and being a "person" and "citizen" for so many statutory and constitutional purposes, that we forget how perplexing the notion was to early jurists. "That invisible, intangible and artificial being, that mere legal entity," Chief Justice Marshall wrote of the corporation in *Bank of the United States v. Deveaux* – could a suit be brought in its name? Ten years later, in the Dartmouth College case, he was still refusing to let pass unnoticed the wonder of an entity "existing only in contemplation of law." Yet, long before Marshall worried over the personification of the modern corpo-

ration, the best medieval legal scholars had spent hundreds of years struggling with the legal nature of those great public "corporate bodies," the
Church and the State. How could they exist in law, as entities transcending
the living pope and king? It was clear how a king could bind himself – on this
honor – by a treaty. But when the king died, what was it that was burdened
with the obligations of, and claimed the rights under, the treaty his tangible
hand had signed? The medieval mind saw (what we have lost our capacity to
see) how unthinkable it was, and worked out the most elaborate conceits and
fallacies to serve as anthropomorphic flesh for the Universal Church and the
Universal Empire.

It is this note of the unthinkable that I want to dwell upon for a moment.
Throughout legal history, each successive extension of rights to some new
entity has been, theretofore, a bit unthinkable. Every era is inclined to suppose the rightlessness of its rightless "things" to be a decree of Nature, not
a legal convention – an open social choice – acting in support of some status
quo. It is thus that we avoid coming face to face with all the moral, social,
and economic dimensions of what we are doing. Consider, for example, how
the United States Supreme Court sidestepped the moral issues behind slavery in its 1856 *Dred Scott* decision; blacks had been denied to rights of citi-
zenship "as a subordinate and inferior class of beings." Their unfortunate
legal status reflected, in other words, not our choice at all, but "just the way
things were." In an 1856 contest over a will, the deceased's provision that his
slaves should decide between emancipation and public sale was held void on
the ground that slaves had no legal capacity to choose. "These decisions,"
the Virginia court explained,

> are legal conclusions flowing naturally and necessarily from the one
> clear, simple, fundamental idea of chattel slavery. That fundamental
> idea is, that, in the eye of the law, so far certainly as civil rights and rela-
> tions are concerned, the slave is not a person, but a thing. The investi-
> ture of a chattel with civil rights or legal capacity is indeed a legal sole-
> cism and absurdity. The attribution of a legal conscience, legal intellect,
> legal freedom, or liberty and power of free choice and action, and cor-
> responding legal obligations growing out of such qualities, faculties and
> action – implies a palpable contradiction in terms.

In a like vein, the highest court in California once explained that Chinese
had not the right to testify against white men in criminal matters because
they were "a race of people whom nature has marked as inferior, and who
are incapable of progress or intellectual development beyond a certain

point . . . between whom and ourselves nature has placed an impassable difference."

The popular conception of the Jew in the thirteenth century contributed to a law which treated them, as one legal commentator has observed, as "men *ferae naturae,* protected by a quasi-forest law. Like the roe and the deer, they form an order apart." Recall, too, that it was not so long ago that the fetus was "like the roe and the deer." In an early suit attempting to establish a wrongful death action on behalf of a negligently killed fetus (now widely accepted practice in American courts), Holmes, then on the Massachusetts Supreme Court, seems to have thought it simply inconceivable "that a man might owe a civil duty and incur a conditional prospective liability in tort to one not yet in being." The first woman in Wisconsin who thought she might have a right to practice law was told that she did not. We had nothing against *them,* of course; but they were *naturally* different.

> The law of nature destines and qualifies the female sex for the bearing and nurture of the children of our race and for the custody of the homes of the world . . . lifelong callings of women, inconsistent with these radical and sacred duties of their sex, as is the profession of the law, are departures from the order of nature; and when voluntary, treason against it . . . The peculiar qualities of womanhood, its gentle graces, it quick sensibility, its tender susceptibility, its purity, its delicacy, its emotional impulses, its subordination of hard reasons to sympathetic feeling, are surely not qualifications for forensic strife. Nature has tempered woman as little for the juridical conflicts of the court room, as for the physical conflicts of the battle field . . .

The fact is, that each time there is a movement to confer rights onto some new "entity" the proposal is bound to sound odd or frightening or laughable.[1] This is partly because until the rightless thing receives its rights, we

---

1. Recently, a group of prison inmates in Suffolk County tamed a mouse that they discovered, giving him the name Morris. Discovering Morris, a jailer flushed him down the toilet. The prisoners brought a proceeding against the warden complaining, *inter alia,* that Morris was subjected to discriminatory discharge and was otherwise unequally treated. The action was unsuccessful, the court noting that the inmates themselves were "guilty of imprisoning Morris without a charge, without a trial, and without bail," and that other mice at the prison were not treated more favorably. "As to the true victim, the Court can only offer again the sympathy first proffered to his ancestors by Robert Burns's poem, 'To a Mouse.'"

The whole matter seems humorous, of course. But we need to know more of the function of humor in the unfolding of a culture, and the ways in which it is involved with the social growing pains to which it is testimony. Why do people make jokes about the Women's Liberation Movement? Is it not on account

cannot see if as anything but a thing for the use of "us" – those who are holding rights at the time.

Thus it was that the Founding Fathers would speak of the inalienable rights of all men, and yet maintain a society that was, by modern standards, without the most basic rights for blacks, Indians, children and women. There was no hypocrisy; emotionally, no one felt that these other things were *men.* In this vein, what is striking about the Wisconsin case above is that the court, for all its talk about women, so clearly was never able to see women as they are and might become. All it could see was the popular "idealized" version of an object it needed. Such is the way the slave South looked upon the black. "The older South," W. E. Du Bois wrote, clung to "the sincere and passionate belief that somewhere between men and cattle, God created a *tertium quid,* and called it a Negro."

Obviously, there is something of a seamless web involved: there will be resistance to giving a "thing" rights until it can be seen and valued for itself; yet, it is hard to see it and value a "thing" for itself until we can bring ourselves to give it rights – which is almost inevitably going to sound inconceivable to a large group of people.

The reader must know by now, if only from the title of the book, the reason for this little discourse on the unthinkable. I am quite seriously proposing that we recognize legal rights of forests, oceans, rivers and other so-called "natural objects" in the environment – indeed, of the natural environment as a whole.

As strange as such a notion may sound, it is neither fanciful nor without considerable operational significance. In fact, I do not think it would be a misdescription of recent developments in the law to say that we are already on the verge of such an assignment of rights to nature, although we have not faced up to what we are doing in those particular terms.

We should do so now, and begin to explore the implications such an idea would yield.

---

of – rather than in spite of – the underlying validity of the protests and the uneasy awareness that a recognition of the claims is inevitable?

Sierra Club vs. Morton, Secretary of the Interior

Despite Christopher Stone's arguments, the Supreme court of the United States found against the Sierra Club. Below is a part of the final opinion, written by Justice Stewart, as well as the dissenting opinion written by Justice Douglas, who clearly had been influenced by Stone's work.

These opinions are reproduced in an excellent book, *Legal and Ethical Concepts in Engineering* by Keith Blinn (Prentice-Hall, Englewood Cliffs NJ, 1989), and are used to illustrate the how the judicial system manages such sticky questions as the standing of natural object. The book is exceptionally useful in assisting the reader in the understanding of this and other legal problems faced by engineers.

Mr. Justice Stewart: [*Editors' note:* Following a review of the facts of the case, Justice Stewart addresses the "standing" issue.] The first question presented is whether the Sierra Club has alleged facts that entitle it to obtain judicial review of the challenged action. Whether a party has a sufficient stake in an otherwise justiciable controversy to obtain judicial resolution of that controversy is what has traditionally been referred to as the question of standing to sue. Where the party does not rely on any specific statute authorizing invocation of the judicial process, the question of standing depends upon whether the party has alleged such a "personal stake in the outcome of the controversy," as to ensure that "the dispute sought to be adjudicated will be presented in an adversary context and in a form historically viewed as capable of judicial resolution." Where, however, Congress has authorized public officials to perform certain functions according to law, and has provided by statute for judicial review of those actions under certain circum-

Supreme Court transcript 405 U.S. 727, 92 S. Ct. 1361 (1972).

stances, the injury as to standing must begin with a determination of whether the statute in question authorizes review at the behest of the plaintiff.

The Sierra Club relies upon Section 10 of the Administrative Procedure Act (APA) . . . which provides:

> A person suffering legal wrong because of agency action; or adversely affected or aggrieved by agency action within the meaning of a relevant statute, is entitled to judicial review thereof.

Early decisions under this statute interpreted the language as adopting the various formulations of "legal interest" and "legal wrong" then prevailing as constitutional requirements of standing.

The injury alleged by the Sierra Club will be incurred entirely by reason of the change in the uses to which Mineral King will be put, and the attendant change in the aesthetics and ecology of the area. Thus, in referring to the road to be built through Sequoia National Park, the complaint alleged that the development "would destroy or otherwise adversely affect the scenery, natural and historic objects and wildlife in the park and would impair the enjoyment of the park for future generations." We do not question that this type of harm may amount to an "injury in fact" sufficient to lay the basis for standing under Section 10 of the APA. Aesthetic and environmental well-being, like economic well-being, are important ingredients of the quality of life in our society, and the fact that particular environmental interests are shared by the many rather than the few does not make them less deserving of legal protection through the judicial process. But the "injury in fact" test requires more than an injury to a cognizable interest. It requires that the party seeking review be himself among the injured.

The impact of the proposed changes in the environment of Mineral King will not fall indiscriminately upon every citizen. The alleged injury will be felt directly only by those who use Mineral King and Sequoia National Park, and for whom the aesthetic and recreational values of the area will be lessened by the highway and ski resort. The Sierra Club failed to allege that it or its members would be affected in any of their activities or pastimes by the Disney development. Nowhere in the pleadings or affidavits did the Club state that its members use Mineral King for any purpose, much less that they use it in any way that would be significantly affected by the proposed actions of the respondents.

The Club apparently regarded any allegations of individualized injury as superfluous, on the theory that this was a "public" action involving questions as to the use of natural resources, and that the Club's long-standing concern with and expertise in such matters were sufficient to give it standing as a "representative of the public." This theory reflects a misunderstanding of our cases involving so-called "public actions" in the area of administrative law.

It is clear that an organization whose members are injured may represent those members in a proceeding for judicial review. But a mere "interest in a problem," no matter how long-standing the interest and no matter how qualified the organization is in evaluating the problem, is not sufficient by itself to render the organization "adversely affected" or "aggrieved" within the meaning of the APA. The Sierra Club is a large and long-established organization, with a commitment to the cause of protecting our Nation's natural heritage from man's depredations. But if a "special interest" in this subject were enough to entitle the Sierra Club to commence this litigation, there would appear to be no objective basis upon which to disallow a suit by any other bona fide "special interest" organization, however small or short-lived. And if any group with a bona fide "special interest" could initiate such litigation, it is difficult to perceive why any individual citizen with the same bona fide special interest would not also be entitled to do so.

The requirement that a party seeking review must allege facts showing that he is himself adversely affected does not insulate executive action from judicial review, nor does it prevent any public interests from being protected through the judicial process. It does serve as at least a rough attempt to put the decision as to whether review will be sought in the hands of those who have a direct stake in the outcome. That goal would be undermined were we to construe the APA to authorize judicial review at the behest of organizations or individuals who seek to do no more than vindicate their own value preferences through the judicial process.

The principle that the Sierra Club would have us establish in this case would do just that.

[W]e conclude that the Court of Appeals was correct in holding that the Sierra Club lacked standing to maintain this action . . . The judgement is affirmed.

Mr. Justice Douglas, dissenting: I share the view of my Brother Blackmun and would reverse the judgment below.

The critical question of "standing" would be simplified and also put neatly in focus if we fashioned a federal rule that allowed environmental issues to be litigated before federal agencies or federal courts in the name of the inanimate object about to be despoiled, defaced, or invaded by roads and bulldozers and where injury is to the subject of public outrage. Contemporary public concern for protecting nature's ecological equilibrium should lead to the conferral of standing upon environmental objects to sue for their own preservation. See Stone, *Should Trees Have Standing? – Toward Legal Rights for Natural Objects,* 45 S. Cal. L. Rev. 450 (1972). This suit would therefore be more properly labeled as *Mineral King v. Morton.*

Inanimate objects are sometimes parties in litigation. A ship has a legal personality, a fiction found useful for maritime purposes. The corporation sole, a creature of ecclesiastical law, is an acceptable adversary and large fortunes ride on its cases. The ordinary corporation is a "person for purposes of the adjudication processes, whether it represents proprietary spiritual, aesthetic or charitable causes."

So it should be as respects valleys, alpine meadows, rivers, lakes, estuaries, beaches, rivers, groves of trees, swampland, or even air that feels the destructive pressures of modern technology and modern life. The river, for example, is the living symbol of all the life it sustains or nourishes – fish, aquatic insects, water fowls, otter, deer, elk, bear, and all other animals, including man, who are dependent on it or who enjoy it for its sight, its sound, or its life. The river as plaintiff speaks for the ecological unit of life that is part of it. Those people who have a meaningful relation to that body of water – whether it be fisherman, a canoeist, a zoologist, or a logger – must be able to speak for the values which the river represents and which are threatened with destruction.

I do not know Mineral King. I have never seen it nor traveled it, though I have seen articles describing its proposed "development" notably Hano, "Protectionists vs. Recreationists – The Battle of Mineral King," *N. Y. Times Mag.,* Aug. 17, 1969, p. 25. The Sierra Club in its complaint alleges that "[o]ne of the principal purposes of the Sierra Club is to protect and conserve the national resources of the Sierra Nevada Mountains." The District Court held that this uncontested allegation made the Sierra Club "sufficiently aggrieved" to have "standing" to sue on behalf of Mineral King.

Mineral King is doubtless like other wonders of the Sierra Nevada such as Tuolone Meadows and the John Muir trail. Those who hike it, fish it, hunt it, camp in it, frequent it, or visit it merely to sit in solitude and wonderment are legitimate spokesmen for it, whether they may be few or many. Those who have that intimate relation with the inanimate object about to be injured, polluted, or otherwise despoiled are its legitimate spokesmen.

The Solicitor, whose views on this subject are in the Appendix to this opinion, takes a wholly different approach. He considers the problem in terms of "government by the Judiciary." With all respect, the problem is to make certain that the inanimate objects, which are the very core of America's beauty, have spokesmen before they are destroyed.

Yet the pressures on agencies for favorable action one way or the other are enormous. The suggestion that Congress can stop action which is undesirable is true in theory; yet even Congress is too remote to give meaningful direction and its machinery is too ponderous to use very often. The federal agencies of which I speak are not venal or corrupt. But they are notoriously under the control of powerful interests who manipulate them through advisory committees, or friendly working relations, or who have that natural affinity with the agency which in time develops between the regulator and the regulated.

The Forest Service – one of the federal agencies behind the scheme to despoil Mineral King – has been notorious for its alignment with lumber companies, although its mandate from Congress directs it to consider the various aspects of multiple use in its supervision of the national forests.

The voice of the inanimate object, therefore, should not be stilled. That does not mean that the judiciary takes over the managerial functions from the federal agency. It merely means that before these priceless bits of Americana (such as a valley, an alpine meadow, a river, or a lake) are forever lost or are so transformed as to be reduced to the eventual rubble of our urban environment, the voice of the existing beneficiaries of these environmental wonders should be heard.

Perhaps they will not win. Perhaps the bulldozers of progress will plow under all the aesthetic wonders of this beautiful land. That is not the present question. The sole question is, who has the standing to be heard?

The Rights of Natural Objects

KRISTIN SHRADER-FRECHETTE

Concluding the battle over the rights of natural things, Shrader-Frechette uses philosophical instead of legal arguments for the rights of natural objects. This excerpt is reproduced from her book *Environmental Ethics*.

A number of ethical and legal considerations suggest that there is a strong rational basis for affirming that natural objects have rights. According to Professor Stone, our concept of "legal right" is a product of traditional, historical mental baggage, rather than the result of clear analysis. Since the future of the planet is dependent upon humanity's adopting a less homocentric view of legal rights, Stone proposes that full legal rights be extended to the inanimate world. Natural objects would be allowed to sue for damages, to have their injuries considered in measuring relief, and to recover damages for their benefit. Under this proposal, a friend of the natural object in question could apply to the court for guardianship, just as a friend may now do on behalf of a legal incompetent. Since the guardian could sue to protect the rights of the natural object, any damages awarded could be placed in a trust fund to be administered by the guardian on behalf of the natural object.

In building his case for the thesis that nature ought to be accorded rights, Stone points out (as did Leopold) that historically, children, women, and aged, prisoners, aliens, Blacks, Indians, and the insane were treated as objects or as things, rather than as persons possessing rights.[1] Hence, he argues, it is not unthinkable to recognize the rights of natural objects, both because legal history reveals that rights are continually being expanded in

Excerpted from K. S. Shrader-Frechette, *Environmental Ethics* (The Boxwood Press, Pacific Grove, CA, 1981). Used with permission.

1.   C. Stone, *Should Trees Have Standing?* (William Kaufmann, Los Altos, CA, 1974).

principle and in practice, and because each time rights are conferred on some new entity, the proposal sounds odd, frightening, or laughable.

One of the most common arguments, that the environment has rights, is based on the fact that society already recognizes the existence of numerous inanimate rights-holders. These include trusts, corporations, municipalities, and ships among others. To recognize the rights of nature, however, means neither that inanimate objects would have the same rights as human beings, nor that everything in the environment would have the same rights as every other natural object. Rather, to say that nature has legal rights simply means that legal actions may be instituted at its behest, that injury to it must be taken into account by the courts, and that relief may be given to its benefit.

At present natural objects have no such legal rights. Nature has no "standing" to sue for damages, for example, at its own behest; only humans able to show possible invasions of their rights have standing. Moreover, even when someone is willing to establish standing, the merits of a case (e.g., involving pollution of a natural object such as a stream) are currently decided on the basis of the conflicting interests of the (human) plaintiff and defendant. The fate of the natural object, *per se,* is ignored. In cases involving injury to the environment, the beneficiary of a favorable judgment is always a human. No money goes to the benefit of the natural object to repair its damages.

Such a procedure, it can be argued, is not wise because it does not enable nature to seek redress in its own behalf. Just as legal incompetents have guardians to manage their affairs, so also friends of natural objects could be recognized by the courts as their guardians. In this way, the Environmental Defense Fund, for example, might raise the land's rights in the land's name. One reason for doing so might be to prevent unsound strip-mining operations.

Besides the fact that there are already inanimate rights-holders, another argument for establishing the legal rights of *nature* is that such recognition will further enable society to become aware of all the costs of its activities. If the legal rights of the environment are not recognized, for example, in the case of a stream polluted by a paper mill, then there is no feasible way for all the fragmented and unrepresented damage claims to be brought before the courts. For practical reasons, the claims will not be pressed. Owners of summer homes and sellers of fishing bait, for example, are unlikely to bring forward their claims. These distantly injured humans, including unborn generations, however, could all be represented by the guardian of the natural object, if the object itself were made the focus of damages to it.

If injury to the environment were viewed as a violation of rights, then it would be a simple matter for a guardian to prove an economically measurable loss in court. A calculation of the damages to a natural object might be thought of as analogous to a measure of the injuries to an automobile-accident victim. In both cases, the responsible party might bear the expense of making either the person or the natural object "whole" or healthy.[2]

A third reason for recognizing the rights of natural objects is based on the recognition that we share fundamental needs with the nonhuman world. Plants and humans, for example, both require water, oxygen, and nutrition; both grow, reproduce, and die. With these bases for empathy and identity established, human moral evolution could conceivably result in the extension of rights to the nonhuman world. "The inner dynamic of every assault on domination is an ever broadening realization of reciprocity and identity."[3]

If we humans realized the reciprocity in our relationships with the nonhuman world, and if we identified and empathized with other beings on the planet, then we would have to give up some psychic investment in humankind's sense of separateness and specialness. Although this might be difficult, it could also be desirable, in enabling us to "free ourselves of needs for supportive illusions."[4] Extending legal rights to natural objects thus might enable us to reaffirm several truths, viz., "that the oppressor is among the first to be liberated when he lifts the yoke," and "that freedom can be realized only in fidelity to obligation."[5] In other words, recognizing the legal rights of nature might develop not only in humanity's capacity for empathy and understanding but also its ability to live free from the illusion that members of only one species are the center of the universe.

2. See ibid.
3. L. H. Tribe, "Ways to Think About Plastic Trees: New Foundations for Environmental Law," *Environmental Law Review – 1975* (Clark Boardman, New York, 1975).
4. See Stone.
5. See Tribe.

Supplemental Readings for Chapter 6

Ecological Feminism

KAREN J. WARREN

One most controversial and yet most interesting approaches to environmental ethics is ecofeminism, a word first coined by Françoise d'Eaubonne in 1974. A parallel is claimed between the domination of nature and the domination of women. Karen Warren goes further than that in this excerpt from her essay, arguing that no environmental ethic is worth anything unless it incorporates the feminist ideals into its philosophy. Karen Warren teaches in the philosophy department at Macalaster College.

A feminist ethic involves a twofold commitment to critique male bias in ethics whenever it occurs, and to develop ethics which are not male-biased. Sometimes this involves articulation of values (e.g., values of care, appropriate trust, kinship, friendship) often lost or underplayed in mainstream ethics. Sometimes it involves engaging in theory building by pioneering in new directions or by revamping old theories in gender sensitive ways. What makes the critique of old theories or conceptualizations of new ones "feminist" is that they emerge out of sex-gender analyses and reflect whatever those analyses reveal about gendered experience and gendered social reality.

As I conceive feminist ethics in the pre-feminist present, it rejects attempts to conceive of ethical theory in terms of necessary and sufficient conditions, because it assumes that there is no essence (in the sense of some transhistorical, universal, absolute abstraction) of feminist ethics. While attempts to formulate joint necessary and sufficient conditions of a feminist ethic are unfruitful, nonetheless, there are some necessary conditions, what I prefer to call "boundary conditions," of a feminist ethic. These boundary

Excerpted from K. Warren, "The Power and Promise of Ecological Feminism," *Environmental Ethics*, vol. 12, no. 2 (1990). Used with permission.

conditions clarify some of the minimal conditions of a feminist ethic without suggesting that feminist ethics has some ahistorical essence. They are like the boundaries of a quilt or collage. They delimit the territory of the piece without dictating what the interior, the design, the actual pattern of the piece looks like. Because the actual design of the quilt emerges from the multiplicity of voices of women in a cross-cultural context, the design will change over time. It is not something static.

What are some of the boundary conditions of a feminist ethic? First, nothing can become part of a feminist ethic – can be part of the quilt – that promotes sexism, racism, classism, or any other "isms" of social domination. Of course, people may disagree about what counts as a sexist act, racist attitude, classist behavior. What counts as sexism, racism, or classism may vary cross-culturally. Still, because a feminist ethic aims at eliminating sexism and sexist bias, and sexism is intimately connected in conceptualization and in practice to racism, classism, and naturism, a feminist ethic must be anti-sexist, anti-racist, anti-classist, anti-naturist and opposed to any "ism" which presupposes or advances a logic of domination.

Second, a feminist ethic is a *contextualist* ethic. A contextualist ethic is one which sees ethical discourse and practice as emerging from the voices of people located in different historical circumstances. A contextualist ethic is properly viewed as a *collage* or *mosaic*, a *tapestry* of voices that emerges out of felt experiences. Like any collage or mosaic, the point is not to have *one picture* based on a unit of voices, but a *pattern* which emerges out of the very different voices of people located in different circumstances. When a contextualist ethic is *feminist*, it gives central place to the voices of women.

Third, since a feminist ethic gives central significance to the diversity of women's voices, a feminist ethic must be structurally pluralistic rather than unitary or reductionistic. It rejects the assumption that there is "one voice" in terms of which ethical values, beliefs, attitudes, and conduct can be assessed.

Fourth, a feminist ethic reconceives ethical theory as theory in process which will change over time. Like all theory, a feminist ethic is based on some generalizations. Nevertheless, the generalizations associated with it are themselves a pattern of voices within which the different voices emerging out of concrete and alternative descriptions of ethical situations have meaning. The coherence of a feminist theory so conceived is given within a historical and conceptual context, i.e., within a set of historical, socioeconomic circumstances (including circumstances of race, class, age, and affectual orientation) and within a set of basic beliefs, values, attitudes, and assumptions about the world.

Fifth, because a feminist ethic is contextualist, structurally pluralistic, and "in-progress," one may evaluate the claims of a feminist ethic in terms of their *inclusiveness;* those claims (voices, patterns of voices) are morally and epistemologically favored (preferred, better, less partisan, less biased) which are more inclusive of the felt experiences and perspectives of oppressed persons. The condition of inclusiveness requires and ensures that the diverse voices of women (as oppressed persons) will be given legitimacy in ethical theory building. It thereby helps to minimize empirical bias. e.g., bias rising from faulty or false generalizations based on stereotyping, too small a sample size, or a skewed sample. It does so by ensuring that any generalizations which are made about ethics and ethical decision making include – indeed cohere with – the pattered voices of women.

Sixth, a feminist ethic makes no attempt to provide an "objective" point of view, since it assumes that in contemporary culture there really is no such point of view. As such, it does not claim to be "unbiased" in the sense of "value-neutral" or "objective." However, it does assume that whatever bias it has as an ethic centralizing the voices of oppressed persons is a *better bias* – "better" because it is more inclusive and therefore less partial – than those which exclude those voices.

Seventh, a feminist ethic provides a central place for values typically unnoticed, underplayed, or misrepresented in traditional ethics, e.g., values of care, love, friendship, and appropriate trust. Again, it need not do this at the exclusion of considerations of rights, rules, or utility. There may be many contexts in which talk of rights or of utility is useful or appropriate, for instance, in contracts or property relationships, talk of rights may be useful and appropriate. In deciding what is cost-effective or advantageous to the most people, talk of utility may be useful and appropriate. In a feminist *qua* contextualist ethic, whether or not such talk is useful or appropriate depends on the context; *other values* (e.g. values of care, trust, friendship) are *not* viewed as reducible to or captured solely in terms of such talk.

Eighth, a feminist ethic also involves a reconception of what it is to be human and what it is for humans to engage in ethical decision making, since it rejects as either meaningless or currently untenable gender-free or gender-neutral description of humans, ethics, and ethical decision making. It thereby rejects what Alison Jaggar calls "abstract individualism," i.e., the position that it is possible to identify a human essence or human nature that exists independently of any particular historical context. Humans and human moral conduct are properly understood essentially (and not merely accidentally) in terms of network or webs of historical and concrete relationships.

All the props are now in place for seeing how ecofeminism provides the framework for a distinctly feminist and environmental ethic. It is a feminism that critiques male bias wherever it occurs in ethics (including environmental ethics) and aims at providing an ethic (including an environmental ethic) which is not male biased – and it does so in a way that satisfies the preliminary boundary conditions of a feminist ethic.

First, ecofeminism is quintessentially anti-naturist. Its anti-naturism consist in the rejection of any way of thinking about or acting toward nonhuman nature that reflects a logic, values, or attitude of domination. Its anti-naturist, anti-sexist, anti-racist, anti-classist (and so forth, for all other "isms" of social domination) stance forms the outer boundary of the quilt: nothing gets on the quilt which is naturist, sexist, racist, classist, and so forth.

Second, ecofeminism is a contextualist ethic. It involves a shift *from* a conception of ethics as primarily a matter of rights, rules or principles predetermined and applied in specific cases to entities viewed as competitors in the contest of moral standing, *to* a conception of ethics as growing out of what Jim Cheney calls "defining relationships," i.e., relationships conceived in some sense as defining who one is. As a contextualist ethic, it is not that rights, or rules, or principles are *not* relevant or important. Clearly they are in certain contexts and for certain purposes. It is just that what *makes* them relevant or important is that those to whom they apply are entities *in relationship with* others.

Ecofeminism also involves an ethical shift *from* granting moral consideration to nonhumans *exclusively* on the grounds of some similarity they share with humans (e.g., rationality, interests, moral agency, sentience, rightholder status) *to* "a highly contextual account to see clearly what a human being is and what the nonhuman world might be, morally speaking, *for* human beings."[1] For an ecofeminist, *how* a moral agent is in relationship to another becomes a central significance, not simply *that* a moral agent is amoral agent or is bound by rights, duties, virtue, or utility to act in a certain way.

Third, ecofeminism is structurally pluralistic in that is presupposes and

---

1.   Jim Cheney, "Eco-Feminism and Deep Ecology," *Environmental Ethics*, vol. 9, no. 2 (1987), p. 144.

maintains difference – difference among humans as well as between humans and at least some elements of nonhuman nature. Thus, while ecofeminism denies the "nature/culture" split, it affirms that humans are both members of an ecological community (in some respects) and different from it (in other respects). Ecofeminism's attention to relationships and community is not, therefore, erasure of difference but a respectful acknowledgement of it.

Fourth, ecofeminism reconceives theory in process. It focuses on patterns of meaning which emerge, for instance, from the storytelling and first-person narratives of women (and others) who deplore the twin dominations of women and nature. The use of narrative is one way to ensure that the content of the ethic – the pattern of the quilt – may/will change over time as the historical and material realities of women's lives change and as more is learned about women-nature connections and the destruction of the nonhuman world.

Fifth, ecofeminism is inclusivist. It emerges from the voices of women who experience the harmful dominance of nature and the way that domination is tied to their domination as women. It emerges from listening to the voices of indigenous people such as Native Americans who have been dislocated from their land and have witnessed the attendant undermining of such values as appropriate reciprocity, sharing and kinship that characterize traditional Indian culture. It emerges from listening to voices of those who, like Nathan Hare, critique traditional approaches to environmental ethics as white and bourgeois, and as failing to address issues of "black ecology" and the "ecology" of the inner city and urban spaces. It also emerges out of the voices of Chipko women who see the destruction of "earth, soil and water" as intimately connected with their own inability to survive economically. With its emphasis on inclusivity and difference, ecofeminism provides a framework for recognizing that what counts as ecology and what counts as appropriate conduct toward human and nonhuman environments is largely a matter of context.

Sixth, as a feminism, ecofeminism makes no attempt to provide an "objective" point of view. It is a social ecology. It recognizes the twin dominations of women and nature as problems rooted both in very concrete, historical, socioeconomic circumstances and in oppressive patriarchal conceptual frameworks which maintain and sanction these circumstances.

Seventh, ecofeminism makes a central place for values of care, love, friendship, trust, and appropriate reciprocity – values that presuppose that our relationships to theirs are central to our understanding of who we are. It thereby gives voice to the sensitivity that in climbing a mountain, one is doing something in relationship with an "other," an "other" whom one can come to care about and treat respectfully.

Lastly, an ecofeminist ethic involves a reconception of what it means to be human, and in what human ethical behavior consists. Ecofeminism denies abstract individualism. Humans are who we are in large part by virtue of the historical and social contexts and the relationships we are in, including our relationships with nonhuman nature. Relationships are not something extrinsic to who we are, not an "add on" feature of human nature; they play an essential role in shaping what it is to be human. Relationships of humans to the nonhuman environment are, in part, constitutive of what it is to be human.

By making visible the interconnections among the dominations of women and nature, ecofeminism shows that both are feminist issues and that explicit acknowledgement of both is vital to any responsible environmental ethic. Feminism *must* embrace ecological feminism if it is to end the domination of women because the domination of women is tied conceptually and histori-cally to the domination of nature.

A responsible environmental ethic also *must* embrace feminism. Other-wise, even the seemingly more revolutionary, liberational, and holistic eco-logical ethic will fail to take seriously the interconnected dominations of nature and women that are so much a part of the historical legacy and con-ceptual framework that sanctions the exploitation of nonhuman nature. Failure to make visible these interconnected, twin dominations results in an inaccurate account of how it is that nature has been and continues to be dominated and exploited and produces an environmental ethic that lacks the depth necessary to be truly *inclusive* of the realities of persons who at least in dominant Western cultures have been intimately tied with that exploita-tion, viz., women. Whatever else can be said in favor of such holistic ethics, a failure to make visible ecofeminist insights into the common denominators of the twin oppressions of women and nature is to perpetuate, rather than overcome, the source of that oppression.

The Judeo-Christian Stewardship Attitude Toward Nature

PATRICK DOBEL

Patrick Dobel, at the Graduate School of Public Affairs at the University of Washington in Seattle, begins his essay by noting the number of times he has seen articles on environmental ethics in which the only mention of Judeo-Christian traditions has been the essay by Lynn White, in which White contends that Christianity is the root cause of our environmental problems. Dobel takes issue with this suggestion, and argues in this essay how a clear environmental ethic emerges from the Bible.

The first question to address is the status of the earth and its resources. A different way of putting this is "Who owns the earth?" The answer of the entire Judeo-Christian tradition is clear: God. "In the beginning God created the heavens and the earth" (Gen. 1:3). In direct ethical terms God created the earth, and in distributive justice terms it belongs to him. "The earth is the Lord's and the fullness therefor" (Ps. 24:1). As an act of pure love he created a world and he "founded the earth to endure" (Ps. 119:90–91).

What kind of world did God create? The answer has two dimensions: the physical of descriptive and the ethical. As a product of nature the world was created as a law-bound entity. The laws are derivative of God's will for all creation as "maintained by your rulings" (Ps. 119:90–91). Things coexist in intricate and regulated harmony – the basic postulates of science, mythology and reason. Although we have a world of flaws, it is also a world of bounty and harmony. For it had been promised that "while the earth remains, seed-time and harvest shall not cease" (Gen. 8:22). It was arranged "in wisdom" so that in the balance of nature, "All creatures depend upon you to feed them . . . you provide the food with a generous hand." God's presence ultimately "holds all things in unity" (Col. 1:16–20) and constantly "renews" the world (Ps. 104:24–30). This world abounds in life and is held together in a seamless web maintained by God-willed laws.

In ethical terms, God says that the world was "very good" (Gen. 1:31). In love and freedom he created the world and valued it as good. All the creatures of the world also share in this goodness (I Tim. 4:4). This does not mean that the world is "good for" some purpose or simply has utilitarian value to humanity. The world, in its bounty and multiplicity of life, is independently good and ought to be respected as such.

As an independent good, the earth possesses an autonomous status as an ethical and covenated entity. In Genesis 9:8–17, God directly includes the earth and all the animals as participants in the covenant. He urges the animals to "be fruitful and multiply." Earlier in Genesis 1:30 he takes care specifically to grant the plant life of the earth to the creatures who possess "breath of life." In the great covenant with Noah and all humanity, he expressly includes all other creatures and the earth.

> And God said, "this is sign of the covenant which I make between *me and you and every living creature* that is with you, for all future generations; I set my bow in the sky, and it shall be a sign of the *covenant between* me and the earth. [emphasis added by the author]

The prophets, Isaiah especially, constantly address the earth and describe its independent travail. Paul describes the turmoil and travail of the earth as a midwife of all creation and redemption (Rom. 8:18–20). The earth must be regarded as an autonomous ethical entity bound not just by the restraints of physical law but also by respect for its inherent goodness and the covenant limitations placed upon our sojourn. Perhaps we must think seriously of defining a category of "sins against the earth."

The proper relation between humanity and the bountiful earth is more complex. One fact is of outstanding moral relevance: the earth does not belong to humanity; it belongs to God. Jeremiah summarizes it quite succinctly: "I by my great power and outstretched arm made the earth, land and animals that are on the earth. And I can give them to whom I please" (Jer. 27:5). For an ecological ethic this fact cannot be ignored. The resources and environment of the earth are not ours in any sovereign or unlimited sense; they belong to someone else.

Humanity's relation to the earth is dominated by the next fact: God "bestows" the earth upon all of humanity (Ps. 115:16). This gift does not, however, grant sovereign control. The prophets constantly remind us that God is still the "king" and the ruler/owner to whom the earth revert. No one generation of people possesses the earth. The earth was made "to endure" and was given for all future generations. Consequently the texts constantly reaffirm that the gift comes under covenanted conditions, and that the covenant is "forever." The Bible is permeated with a careful concern for preserving the "land" and the "earth" as an "allotted heritage" (Ps. 2:7–12).

The point is central to the Judeo-Christian response to the world. The world is given to all. Its heritage is something of enduring value designed to benefit all future generations. Those who receive such a gift and benefit from it are duty-bound to conserve the resources and pass them on for future generations to enjoy. An "earth of abundance" (Judg. 18:10) provides for humanity's needs and survival (Gen. 1:26–28, 9:2–5). But the injunction "obey the covenant" (I Chron. 16:14–18) accompanies this gift.

There are some fairly clear principles that direct our covenanted responsibilities toward the earth. Each generation exists only as "sojourner" or "pilgrim." We hold the resources and the earth as a "trust" for future generations. Our covenated relations to the earth – and for that matter, to all human beings – must be predicated upon the recognition and acceptance of the limits of reality. For there is a "limit upon all perfection"(Ps. 119:96) and we must discover and respect the limits upon ourselves, our use of resources, our consumption, our treatment of others and the environment with its delicate ecosystems. Abiding by the covenant means abiding by the laws of nature, both scientific and moral. In ecological terms the balance of nature embodies God's careful plan that the earth and its bounty shall provide for the needs and survival of all humanity of all generations.

The Sanctity of Life in Hinduism

The Hindu religion is ancient by Western standards, and is still practiced by millions of people. Professor Dwivedi, who is on the faculty of the Department of Political Science at the University of Guelph in Canada, argues that this has been no accident. The Hindu religion has within its fabric a strict environmental ethic in which all animals and life are sacred. It is also true, of course, that the Hindu religion has not prevented the more recent environmental destruction of vast stretches of the world. Ideally, however, the basic tenets of Hinduism as they apply to the environment are indeed worth contemplating and even emulating.

The principle of the sanctity of life is clearly ingrained in the Hindu religion. Only God has absolute sovereignty over all creatures, thus, human beings have no dominion over their own lives or nonhuman life. Consequently, humanity cannot act as a viceroy of God over the planet, nor assign degrees of relative worth to other species. The idea of the Divine Being as the one underlying power of unity is beautifully expressed in the Yajurveda:

> The loving sage beholds that Being, hidden mystery, wherein the universe comes to have one home; Therein unites and therefrom emanates the whole; The Omnipresent One pervades souls and matter like warp and woof in created beings. (Yajurveda 32.8).

The sacredness of God's creation means no damage may be inflicted on other species without adequate justification. Therefore, all lives, human and nonhuman, are of equal value and all have the same right to existence.

Excerpted from O. P. Dwivedi, "Satyapraha for Conservation: Awakening the Spirit of Hinduism," in J. R. Engel and J. G. Engel, eds., *Ethics of Environment and Development* (Bellhaven Press, London, 1990). Used with permission.

According to the Atharvaveda, the Earth is not for human beings alone, but for other creatures as well:

> Born of Thee, on thee move mortal creatures; Thou bearest them – the biped and the quadruped; Thine, O Earth, are the five races of men, for whom Surya (Sun), as he rises spreads with his rays the light that is immortal (Atharvaveda 12.1–15). . . .

The most important aspect of Hindu theology pertaining to treatment of animal life is the belief that the Supreme Being was himself incarnated in the form of various species. The Lord says: "This form is the source and indestructible seed of multifarious incarnations within the universe, and from the particle and portion of this form, different living entities, like demigods, animals, human beings and other, are created (*Srimad-Bhagavata* Book I, Discourse III:5). Among the various incarnations of God (numbering from ten to twenty-four depending upon the source of the text), He first incarnated Himself in the form of a fish, then a tortoise, a boar, and a dwarf. His fifth incarnation was a man–lion. As Rama he was closely associated with monkeys, and as Krishna he was always surrounded by the cows. Thus, other species are accorded reverence. . . .

As early as in the time of Regveda, tree worship was quite popular and universal. The tree symbolized the various attitudes of God to the Regvedic seers. Regveda regarded plants as having divine power, with one entire hymn devoted to their praise, chiefly with reference to their healing properties (Regveda 10.97). During the period of the great epics and Purana, the Hindu respect for flora expanded further. Trees were considered as being animate and feeling happiness and sorrow. It is still popularly believed that every tree has a *Vrikksadevata,* or "tree deity," who is worshipped with prayers and offerings of water, flowers, sweets, and encircled with sacred threads.

Islamic Environmental Ethics

MAWIL Y. IZZI DEEN (SAMARRAI)

The religions of Islam holds that people were placed on the earth by God (Allah) and that the entire world is carefully thought out and controlled. This means, according to Professor Mawil Y. Izzi Deen, not that we cannot enjoy ourselves, of course, but that humans must respect the gift we have received. Professor Deen is on the faculty of the King Abdul Aziz University, Jeddah, Saudi Arabia.

In Islam, the conservation of the environment is based on the principle that all the individual components of the environment were created by God, and that all living things were created with different functions, functions carefully measured and meticulously balanced by the Almighty Creator. Although the various components of the natural environment serve humanity as one of their functions, this does not imply that human use is the sole reason for their creation. The comments of the medieval Muslin scholar, Ibn Taymiyah, on those verses of the holy Qur'ān which state that God created the various parts of the environment to serve humanity, are relevant here;

> In considering all these verses it must be remembered that Allah in His wisdom created these creatures for reasons other than serving man, for in these verses he only explains the benefits of these creatures [to man].

The legal and ethical reasons for protecting the environment can be summarized as follows: First, the environment is God's creation and to protect it is to preserve its value as a sign of the Creator. To assume that the envi-

Excerpted from M. Y. I. Deen, "Islamic Environmental Ethics, Law and Society," in J. R. Engel and J. G. Engel, eds., *Ethics of Environment and Development* (Bellhaven Press, London, 1990). Used with permission.

ronment's benefits to human beings are the sole reason for its protection may lead to environmental misuse and destruction.

Second, the component parts of nature are entities in continuous praise of their Creator. Humans may not be able to understand the form or nature of this praise, but the fact that the Qur'ān describes it is an additional reason for environmental preservation:

> The seven heavens and the earth and all that is therein praise Him, and there is not such a thing but hymneth his praise; but ye understand not their praise. Lo! He is ever Clement, Forgiving. (Surah 17:14)

Third, all the laws of nature are laws made by the Creator and based on the concept of the absolute continuity of existence. Although God may sometimes wish otherwise, what happens, happens according to the natural law of God (*sunnah*), and human beings must accept this as the will of the Creator. Attempts to break the law of God must be prevented. As the Qur'ān states:

> Hast thou not seen that unto Allah payeth adoration whosoever is in the heavens and whosoever is in the earth, and the sun, and the moon, and the stars, and the hills, and the trees, and the beasts, and many of mankind (Surah 22:18).

Fourth, the Qur'ān acknowledgement that humankind is not the only community to live in this world – "There is not an animal in the earth, nor a flying creature flying on two wings, but they are peoples like unto you" (Surah 6:38) – means that while humans may currently have the upper hand over other "peoples," these other creatures are beings and, like us, are worthy of respect and protection. The Prophet Muhammad (peace be unto him) considered all living creatures worth of protection (*hurmah*) and kind treatment. He was once asked whether there will be a reward from God for charity shown to animals. His reply was very explicit: "For [charity shown to] each creature which has a wet heart there is a reward." Ibn Hajar comments further upon this tradition, explaining that wetness is an indication of life (and so charity extends to all creatures), although human beings are more worthy of the charity if a choice must be made.

Fifth, Islamic environmental ethics is based on the concept that all human relationships are established on justice (*'adl*) and equity (*ihsahn*): "Lo! Allah enjoineth justice and kindness" (Surah 16:90). The prophetic tradition limits benefits derived at the cost of animal suffering. The Prophet Muhammad instructed: "Verily Allah has prescribed equity (*ihsan*) in all things. Thus if you kill, kill well, and if slaughter, slaughter well. Let each of you sharpen his blade and let him spare suffering to the animal he slaughters."

Sixth, the balance of the universe created by God must also be preserved. For "Everything with Him is measured" (Surah 13:8). Also, "There is not a thing but with Us are the stores thereof. And We send it not down save in appointed measure" (Surah 15:21).

Seventh, the environment is not in the service of the present generation alone. Rather, it is the gift of God to all ages, past, present and future. This can be understood from the general meaning of Surah 2:29, "He it is Who Created for you all that is in the earth." The "you" as used here refers to all persons with no limit as to time or place.

Finally, no other creature is able to perform the task of protecting the environment. God entrusted humans with the duty of viceregency, a duty so onerous and burdensome that no other creature would accept it: "Lo! We offered the trust unto the heavens and the earth and the hills, but they shrank from bearing it and were afraid of it. And man assumed it" (Surah 33:72).

Respect for Nature

GARY SNYDER

Gary Snyder, a poet and member of the 1960s "counterculture," wrote this piece in 1971, anticipating both the "deep ecology" movement and the arguments for the rights of natural objects as expressed by Christopher Stone and later Justice William O. Douglas in the famous the Mineral King case decided before the Supreme Court.

I am a poet. My teachers are other poets, American Indians, and a few Buddhist priests in Japan. The reason I am here is because I wish to bring a voice from the wilderness, by constituency. I wish to be a spokesman for a realm that is not usually represented either in intellectual chambers or in the chambers of government.

I was climbing Glacier Peak in the Cascades of Washington several years ago, on one of the clearest days I had ever seen. When we reached the summit of Glacier Peak we could see almost to the Selkirks in Canada. We could see south far beyond the Columbia River to Mount Hood and Mount Jefferson. And, of course, we could see Mount Adams and Mount Rainier. We could see across Puget Sound to the ranges of the Olympic Mountains. My companion, who is a poet, said: "You mean, there is a senator for all this?"

Unfortunately, there isn't a senator for all that. And I would like to think of a new definition of humanism and a new definition of democracy that would include the nonhuman, that would have representation from those spheres. This is what I think we mean by an ecological conscience.

I don't like Western culture because I think it has much in it that is inherently wrong and that is at the root of the environmental crisis that is not recent; it is very ancient; it has been building up for a millennium. There are

Excepted from G. Snyder, *Turtle Island* (New Directions Publishing Co., New York, 1971). Used with permission.

many things in Western culture that are admirable. But a culture that alien-
ates itself from the very ground of its own being – from the wilderness outside (that is to say, wild nature, the wild, self-contained, self-informing
ecosystems) and from that other wilderness, the wilderness within – is
doomed to a very destructive behavior, ultimately perhaps self-destructive
behavior.

The West is not the only culture that carries these destructive seeds.
China had effectively deforested itself by 1000 A.D. India had effectively
deforested itself by 800 A.D. The soils of the Middle East were ruined even
earlier. The forests that once covered the mountains of Yugoslavia were
stripped to build the Roman fleet, and those mountains have looked like
Utah ever since. The soils of southern Italy and Sicily were ruined by lati-
fundia slave-labor farming in the Roman Empire. The soils of the Atlantic
seaboard in the United States were effectively ruined before the American
Revolution because of the one-crop (tobacco) farming. So the same forces
have been at work in East and West.

You would not think a poet would get involved in these things. But the
voice that speaks to me as a poet, what Westerners have called the Muse, is
the voice of nature herself, whom the ancient poets called the great goddess,
the Magna Mater. I regard that voice as a very real entity. At the root of the
problem where our civilization goes wrong is the mistaken belief that nature
is something less than authentic, that nature is not as alive as man is, or as
intelligent, that in a sense it is dead, and that animals are of so low an order
of intelligence and feeling, we need not take their feelings into account.

A line is drawn between primitive peoples and civilized peoples. I think
there is a wisdom in the worldview of primitive peoples that we have to refer
ourselves to, and learn from. If we are on the verge of postcivilization, then
our next step must take account of the primitive worldview which has traditionally and intelligently tried to open and keep open lines of communication with the forces of nature. You cannot communicate with the forces of
nature in the laboratory. One of the problems is that we simply do not know
much about primitive people and primitive cultures. If we can tentatively
accommodate the possibility that nature has a degree of authenticity and
intelligence that requires that we look at it more sensitively, then we can
move to the next step. "Intelligence" is not really the right word. The ecologist Eugene Odum uses the term "biomass."

Life-biomass, he says, is stored information; living matter is stored information in the cells and in the genes. He believes there is more information

of a higher order of sophistication and complexity stored in a few square yards of forest than there is in all the libraries of mankind. Obviously, that is a different order of information. It is the information of the universe we live in. It is the information that has been flowing for millions of years. In this total information context, man may not be necessarily the highest or most interesting product.

Perhaps one of its most interesting experiments at the point of evolution, if we can talk about evolution in this way, is not man but a high degree of biological diversity and sophistication opening to more and more possibilities. Plants are at the bottom of the food chain; they do the primary energy transformation that makes all the life-forms possible. So perhaps plant-life is what the ancients meant by the great goddess. Since plants support the other life-forms, they became the "people" of the land. And the land – a country – is a region within which the interactions of water, air, and soil and the underlying geology and the overlying (maybe stratospheric) wind conditions all go to create both the microclimates and large climactic patterns that make a whole sphere or realm of life possible. The people in that realm include animals, humans, and a variety of wild life.

What we must find a way to do, then, is incorporate the other people – what the Sioux Indians called the creeping people, and the standing people, and the flying people, and the swimming people – into councils of government. This isn't as difficult as you might think. If we don't do it, they will revolt against us. They will submit non-negotiable demands about our stay on the earth. We are beginning to get non-negotiable demands right now from the air, the water, the soil.

I would like to expand on what I mean by representation here at the Center from these other fields, these other societies, these other communities. Ecologists talk about the ecology of oak communities, or pine communities. They *are* communities. This institute – this Center – is of the order of a kiva of elders. Its function is to maintain and transmit the lore of the tribe on the highest levels. If it were doing its job completely, it would have a cycle of ceremonies geared to its seasons, geared perhaps to the migrations of the fish and to the phases of the moon. It would be able to instruct in what rituals you follow when a child is born, when someone reaches puberty, when someone gets married, when someone dies. But, as you know, in these fragmented times, one council cannot perform all these functions at one time. Still it would be understood that a council of elders, the caretakers of the lore of the culture, would open themselves to representation from other life-forms.

Historically this has been done through art. The painting of bison and bears
in the caves of southern France were of that order. The animals were speaking through the people and making their point. And when, in the dances of
the Pueblo Indians and other peoples, certain individuals became seized, as
it were, by the spirit of the deer, and danced as a deer would dance, or
danced the dance of the corn maidens, or impersonated the squash blossom,
they were no longer speaking for humanity, they were taking it on themselves to interpret, through their humanity, what these other life-forms
were. That is about all we know so far concerning the possibilities of incorporating spokesmanship for the rest of life in our democratic society.

Let me describe how a friend of mine from a Rio Grande pueblo hunts.
He is twenty-seven years old. The Pueblo Indians, and I think probably most
of the other Indians of the Southwest, begin their hunt first, by purifying
themselves. They take emetics, a sweat bath, and perhaps avoid their wife for
a few days. They also try not to think certain thoughts. They go out hunting
in an attitude of humility. They make sure that they need to hunt, that they
are not hunting without necessity. Then they improvise a song while they
are in the mountains. They sing aloud or hum to themselves while they are
walking along. It is a song to the deer, asking the deer to be willing to die for
them. They usually still-hunt, taking a place alongside a trail. The feeling is
that you are not hunting the deer, the deer is coming to you; you make yourself available for the deer that will present itself to you, that has given itself
to you. Then you shoot it. After you shoot it, you cut the head off and place
the head facing east. You sprinkle corn meal in front of the mouth of the
deer, and you pray to the deer, asking it to forgive you for having killed it, to
understand that we all need to eat, and to please make a good report to the
other deer spirits that he has been treated well. One finds this way of handling things and animals in all primitive cultures.

Name Index

Subject Index